Thinking about Thinking

(Or, Do I sincerely want to be right?)

Antony Flew

FONTANA/COLLINS

First published in Fontana 1975
Copyright © Antony Flew 1975

Made and Printed in Great Britain by
William Collins Sons & Co Ltd Glasgow

We must follow the argument wherever it leads.

Socrates
(Fifth Century BC)

Far be from us the dangerous novelty of thinking.

Address to Ferdinand VII of Spain by
the University of Cervera
(Nineteenth century AD)

A moment's thought would have shown him. But a moment is a long time, and thought is a painful process.

A E Housman
(Twentieth century AD)

Many people would sooner die than think. In fact they do.

Bertrand Russell
(Twentieth century AD)

Contents

1 The Basic Equipment

1.1 The first thing to get straight in thinking about thinking is the difference between questions about validity and
questions about truth. But in getting this straight we shall
find that we are also sorting out every other really fundamental notion. For the indispensable notions are all connected. We cannot fully master any one without getting the
same grasp upon the lot. Once the essential preparation is
complete, we may proceed to the main business of the book.
That business is to consider examples of thinking, usually of
bad thinking, in order to learn how to do the job better.
Here and now we have first to clean and tidy the tools.

1.2 The reason to begin precisely where we are beginning
is that thinking about thinking is concerned, at least in the
first instance, with the validity or invalidity of arguments,
rather than with the truth or falsity of propositions. What
is true, or false, is propositions. What is valid, or invalid, is
arguments. These notions, and these distinctions, are
absolutely basic. To say that an argument is true or that a
proposition is valid is as uncomprehending, or as inept, as to
say that someone played scrum half in a soccer match or
scored a lot of tries at cricket.

1.3 Take propositions. There are, of course, propositions
and propositions. Both those mutually advantageous pro-
posals which one businessman makes to another, and those
improper but delightful suggestions which Playboys put to
their intended Playmates, are called quite properly propositions. But in this book—perhaps regrettably—we shall
engage with propositions only in a quite different sense of
the word. In this, our relevant sense, the word ' proposi-

tion ' is defined as, ' whatever may be asserted or denied '. So a proposition for us becomes whatever may be expressed by the that-clause in such sentences as, ' She asserted that he had been there on Wednesday ', or, ' He denied that he had ever met her.'

1.4 In the irrelevant, the proposal sense a proposition may be said to be attractive or unattractive, profitable or unprofitable, and many other things besides. What it cannot either be, or be said to be, is either true or false. In our different sense there are again several things which a proposition may be: demonstrated, for instance, or probable, or refuted. Nevertheless the primary characteristic is truth or falsity. For demonstration here is nothing else but proving that the proposition is true. Refutation again is not merely saying, but showing, that the proposition is false. And the proposition which is probable just is probably true.

1.5 Such propositions are not to be identified with arguments, although all arguments contain propositions. Piety demands that our first example be a dull hack, hallowed by immemorial tradition. Its tedious, trite and trivial character will ensure that no one is distracted from what is being illustrated by any interest in the illustration. Later I shall deploy interesting and important examples. I hope thus to escape the dangers of boring myself and everybody else, or suggesting that the subject itself is as trifling as this first illustration.

1.6 Set out carefully, and piously, the traditional example runs: If *All men are mortal*, and if *Socrates was a man*, then, it follows necessarily, *Socrates was mortal*. This includes three constituent propositions, all printed in italics. The first two are serving here as premises, the last as conclusion. In other contexts and in other arguments what is here conclusion might serve as premise, and what are here premises might be derived as conclusions from other premises.

1.7 The italicization of the constituent propositions, and the representation of the whole argument in a hypothetical form, are both important. The first device brings out two things: first, in general, that arguments are concerned with the logical relations between propositions; and second, in particular, what proposition is being said to be necessarily connected with what two others. Later on much more will be said both about logical relations and about logically necessary connections. For the moment it is sufficient, but necessary, to emphasize that these are always and only relations and connections between propositions.

1.8 The second of the two devices, that of representing a whole argument in hypothetical form, makes it clear why, in order to know whether the exemplary argument in which these three propositions are here embodied is valid, we do not need to know whether any of its constituent propositions is true. We do not for this purpose need to know, because in offering the argument we are not actually saying anything about the truth or falsity of these constituent propositions. It is all hypothetical. Another argument of the same form would be no less valid even if all its three constituent propositions happened in fact to be false. This would be true of the absurd argument: If *All tigers are strictly vegetarian,* and if *Socrates the son of Sophroniscus was a tiger,* then, it follows necessarily, *Socrates the son of Sophroniscus was strictly vegetarian.* As we shall be seeing in Chapter Two such hypothetical deductions, albeit from much more sensible premises, may serve as the initial steps in a more complex pattern of argument. That goes on from the actual falsity of the original conclusion in a valid argument to the further conclusion that at least one of the original premises must also be false.

1.9 Although to say that the present argument is valid is thus not to say that any of the three constituent propositions are true, it does imply the truth of the complex hypothetical proposition: If *All men are mortal,* and if *Socrates was a man,* then, it follows necessarily, *Socrates was mortal.* In asserting this truth what is asserted is that the argument from the two constituent premise propositions to the constituent conclusion proposition is valid. The fact that you can say that the claim that this argument is valid (or invalid) is a true (or false) claim is, however, no more a justification for confounding validity with truth, than the fact that you can say that the contention that this man is a homosexual is a true (or false) contention, is a warrant for identifying homosexuality with truth.

1.10 To say that an argument is deductively valid is, by definition, to say that it would be impossible to assert its premise, or premises, while denying its conclusion, or conclusions, without thereby contradicting yourself. That is what deduction is. We have just seen that an argument may be valid, notwithstanding that both its premise, or premises, and its conclusion, or conclusions, are false. Similarly an argument may be invalid, notwithstanding that both its premise, or premises, and its conclusion, or conclusions, are true.

1.11 Later we shall return to the relations and lack of relations between validity and truth, and I will provide mnemonic illustrations. But the first thing now is to underline the connection between the two concepts of deductive validity and of contradiction, and to explain what is so wrong about contradiction. Suppose someone were to maintain that, although *All men are mortal*, and while of course *Socrates was a man*, still *Socrates was not mortal*. No doubt such apparent irrationality in so simple a case is somewhat hard to imagine. Yet that difficulty should, if anything, make it easier to appreciate that, if only someone did behave so, then we would have to choose between two alternative conjectures. Either he is being in some way disingenuous. Or else he is not fully master of the meanings of all the words which he has uttered.

1.12 On the one hand: perhaps he has some sort of doctrinal commitment to affirm the two premises, while nevertheless equally firmly denying the obvious conclusion. He may want to maintain that Socrates is a man, and as such mortal, and yet that Socrates is a god, and as such not mortal. Certainly there are those who hold that someone, though not Socrates, was at the same time both truly man and truly God. Or maybe our imaginary recalcitrant has his reasons for wanting to say one thing in one context or to one group of people, while saying something altogether inconsistent in another context or to another group of people. This temptation is familiar to us all. It is no prerogative of that scapegoat class, the professional politicians.

1.13 On the other hand: it is also possible that our imaginary someone is careless or confused about the crucial difference between all and some: *Some men are mortal* is consistent, as *All men are mortal* is not, with *Some men are not mortal*. Again there is no call for any far-fetched supposing. We come across all too many cases of people who, having found that something holds in some few cases, proceed forthwith to assume, or even to assert, that the same holds equally in all cases. We have done it ourselves.

1.14 There may appear to be a third possibility, that he is interpreting one of the key terms in one way in one of the premises and in another way in the other premise or in the conclusion. The word ' Socrates ', for instance, might be being employed to refer to one person on one occasion and to another on the other. Again, the first ' mortal ' might be being construed as meaning ' liable to death ', while the second ' not mortal ' was being read as metaphorically ' immortal '—immortal, that is, in that wholly different sense in which a great person who indisputably has died, or will some time die, may nevertheless truly and consistently be numbered among the immortals.

1.15 This apparent third possibility is thus the possibility of equivocation. The word ' equivocation ' is here defined as ' the employment of some word or expression in two or more different senses without distinction in the same context '. If the equivocator realizes that he is equivocating in his employment of one of the key terms, then his performance is certainly disingenuous. If he does not realize this, then he is equally certainly, at least in this context, ' not fully master of the meanings of the words which he has uttered' (1.11).

1.16 The basic point developed in the five previous paragraphs is extremely important. It is none the less so for having been made with a hackneyed traditional example, developed in a somewhat far-fetched way. This basic point is that the terms ' valid ' and ' invalid ' as applied to deductive arguments, and the expression ' deductive argument ' itself, have all to be defined in terms of self-contradiction and the avoidance of self-contradiction. It is because these are thus central notions that our concern with logic inextricably involves us also in concerns with both meaning and truth. The basis of the necessary and inescapable involvement with meaning will be immediately obvious. Given that a valid deductive argument is, by definition, one in which to assert the premises while denying the conclusion is to contradict yourself; then it becomes at once clear that no one can be in a position to know whether or not any argument is valid, except in so far as they are master of the meanings of all its crucial terms.

1.17 It may be more difficult to appreciate that there are necessary connections between logic and truth, and why these connections make it so essential to argue validly and to avoid contradiction. For did not this chapter itself begin by insisting that ' thinking about thinking is concerned, at least in the first instance, with the validity or invalidity of arguments rather than with the truth or falsity of propositions ?' And have we not gone on to assert that arguments may be valid, though both their premises and their conclusions happen to be false; or invalid, though both their premises and their conclusions are in fact true?

1.18 Yes, this was said. It is all true. But it is also true that, though sound argument and a reasonable appreciation of the available evidence may happen sometimes to lead to false conclusions, no man who is indifferent to argument and to evidence can claim to be concerned for truth.

Abraham Lincoln was profoundly right when he wrote, chiding the editor of a Springfield newspaper: ' It is an established maxim and moral that he who makes an assertion without knowing whether it is true or false is guilty of falsehood, and the accidental truth of the assertion does not justify or excuse him.' It is also true that to tolerate contradiction is similarly to be indifferent to truth. For the person who, whether directly or by implication, knowingly both asserts and denies one and the same proposition, shows by that behaviour that he does not care whether he asserts what is false, and not true, or whether he denies what is true, and not false.

1.19 To grasp this point is to raise a perennial personal challenge. Like all such personal challenges it should be seen as being at least as much a challenge to me and to us as it is to you and to them. For whenever and wherever I tolerate self-contradiction, then and there I make it evident, either that I do not care at all about truth, or that at any rate I do care about something else more. It was thus precisely because to affirm the premises of a valid deductive argument, while denying the conclusion, is, by the definition of ' valid deductive argument ', to contradict yourself, that the real Socrates used to demand: ' We must follow the argument wherever it leads.'

1.20 The same personally challenging point, that contradiction must be intolerable to anyone who really cares about truth, can with the help of a little demonstration be made rather more elaborately. Anyone inclined to bridle against ' such logic-chopping ' can without serious loss skip on to paragraph 1.24. The promised, or threatened, demonstration was apparently first mounted in the 1300s of our era, either by Duns Scotus, or by one of his pupils. (It is, by the way, ultimately to the uninhibited polemics of the philosophical opponents of the great Duns Scotus that we all owe our word ' dunce '. There must be some moral here!)

1.21 The demonstration goes like this. First take as your personal premise a contradiction of your choice. I take for mine the conjunction of the two propositions: (1) *The British Labour Party is a socialist party*; and (2) *The British Labour Party is not a socialist party*. Now choose, equally freely, any false proposition. I choose: *W. C. Fields is alive and drunk in Beverly Hills*. Next take one half of the initial contradiction as a separate premise. From *The British Labour Party is a socialist party* it certainly follows that *The British Labour Party is a socialist party and/or W. C. Fields is alive and drunk in Beverly Hills*. For, given that *x is true*, whatever x may be; then, it follows necessarily, for the same value of x and for any value of y, that *x and/or y is true*.

1.22 The only but sufficient justification for employing the symbols ' x ' and ' y '—rather than the awkward verbal alternatives, ' something ', ' the same something ', ' something else ', and ' the same something else '—is that the point can thereby be made more briefly, more clearly, and more elegantly. The object is, as it always should be, to promote understanding. What needs to be understood is that so far it has been shown that, from my arbitrarily chosen contradictory premise, it follows that *The British Labour Party is a socialist party and/or W. C. Fields is alive and drunk in Beverly Hills*. Next we summon the second half of the initial contradiction: (2) *The British Labour Party is not a socialist party*. Taking our previous conclusion as one premise and this as the other, it becomes impossible to avoid the agreeable but false conclusion: *W. C. Fields is alive and drunk in Beverly Hills*. For to deny this, while asserting these two premises, would be to contradict oneself.

1.23 We thus have an absolutely general and absolutely compulsive demonstration that from any contradiction which you like to choose, any other proposition, equally arbitrarily chosen, follows necessarily. By the same token the negation of that other, arbitrarily chosen, proposition must also follow, equally necessarily. We can, that is to say, by the same method also deduce the opposite conclu-

sion: *W. C. Fields is not alive and drunk in Beverly Hills*. Both
every proposition and its negation thus follows from any
contradiction. Hence, if contradiction is tolerated, then, in
a very literal sense, anything goes. This situation must
itself be totally intolerable to anyone who has any concern
at all to know what is in fact true, and to avoid either saying
or implying what is in fact false. If all this seems pedantic,
recall Bertrand Russell's mischievous definition of a pedant:
' A person who prefers his statements to be true.'

1.24 Generally, therefore, when someone with preten-
sions to be a thinker either denounces the restrictions of
logic, or remains unmoved by charges of self-contradiction,
we know what to think. Thomas Aquinas understood as well
as any man that the saints and the prophets may speak of
mysteries. Yet, having a grip on logical fundamentals,
Aquinas never forgot that there can be no place for self-
contradiction in any authentic quest for truth. Thus, in
considering the omnipotence to be attributed to his God, he
took account of what is in modern terms the distinction
between logical and other senses of ' impossibility '. A
suggestion is said to be logically impossible if that suggestion
contains or implies a self-contradiction, or is perhaps otherwise incoherent and unintelligible. But a suggestion which is
not in this sense logically impossible may be ruled out by
the actual laws of nature, and hence be factually impossible.
As Aquinas puts it in the *Summa Theologica*: ' Whatever
does not imply a contradiction is, consequently, among
those possibilities in virtue of which God is described as
omnipotent. But what does imply a contradiction is not
subsumed under the divine omnipotence . . .' (I Q25 A3).
You cannot, he might have said, transmute some incoherent
mixture of words into sense merely by introducing the three-
letter word ' God ' to be its grammatical subject.

1.25 One place where this distinction and this insight is
indispensable is discussion of The Problem of Evil. This is
the theist's problem of trying to show that he is not contra-

dicting himself in maintaining: both that there is, as there is indeed, much evil in the universe; and that that universe is the work of an all-powerful, all-good God. It is not a bit of good to appeal here to what are, the universe being as it happens to be, factual impossibilities. The only hope for the theist is to try to show that it would be logically impossible to have the actual goods without the actual evils; as it is, for instance, logically impossible to have the good of forgiveness without the evil of an injury to be forgiven. It is logically impossible, because it is self-contradictory to speak of forgiving a non-existent injury. But for the theist it must be almost blasphemous to argue here, along lines which I once saw indicated in The Wayside Pulpit: 'If it never rained, there would be no hay to make when the sun shone.'

1.26 Moving from the medieval to the contemporary, we should now know what to think when Herbert Marcuse takes a motto from Horkheimer and Adorno: 'The general concept which discursive logic developed has its foundations in the reality of domination.' According to that Modern Master the prophet Marcuse we have to be emancipated, apparently willy nilly, from the supposedly artificial and oppressive doctrine that 'contradictions are the fault of incorrect thinking' (Marcuse, Chapter 5; compare MacIntyre, pp. 75-79). Certainly it is true that those who speak thus disrespectfully of logic or tolerantly of contradiction usually do so, either because they do not much care whether what they say is true, and or because they have little or no comprehension of what logic is. Yet if we are to perfect our own grasp we need to take some account of various other temptations to these unfortunate ways of talking.

1.27 One major source of present confusion is the persistence in Marxist circles of the Hegelian practice of speaking of contradictions not only in discourse but also in the non-verbal universe around us. Thus Mao Tse-tung wrote in his essay *On contradiction*: 'The supersession of the old by the new is the universal, for ever inviolable law of the world. . . .

Everything contains a contradiction between its new aspect and its old aspect, which constitutes a series of intricate struggles. . . . At the moment when the new aspect has won the dominant position over the old aspect, the quality of the old thing changes into the quality of the new thing. Thus the quality of a thing is mainly determined by the principal aspect of the contradiction that has won the dominant position.'

1.28 A contradiction in this regrettable usage is thus, not a verbal contradiction, but a conflict or a tension in or between things or people. Once these categories are properly distinguished the apparent justification for employing the same word in two utterly different cases disappears; while to the extent that the usage helps to collapse or to confound a categorical distinction it is to be deplored. This same usage encourages talk of fruitful or even non-antagonistic contradictions, contradictions which are welcome, or at least venial. (Mao Tse-tung himself continued, speaking of the ' contradiction ' between town and country: ' But in a socialist nation and in our revolutionary bases such an antagonistic contradiction becomes a non-antagonistic contradiction; and it will disappear when a communist society is realized. . . .')

1.29 But talk of fruitful (if not perhaps of non-antagonistic) contradictions may have quite a different source. The contradictions then referred to are genuine verbal or symbolic self-contradictions, and the fruit offered has to be picked by labouring to remove the contradiction. The vital point for us is that this fruitfulness presupposes the removal of the contradiction. It is only in so far as contradiction is recognized to be intolerable that the labours which provide the fruit have to begin.

1.30 Our hypothetical recalcitrant wanted to maintain: both that Socrates was a man, and as such mortal; and that Socrates was a god, and as such immortal. This appears to be contradictory. But the real or apparent contradiction may become productive if it generates the discovery that, and how, the two seeming incompatibles are after all compatible. Again, if someone were actually to say that the British Labour Party is both a socialist party and not a socialist party, then this contradiction too might prove fertile. But the condition of the fertility must always be that the contradiction is removed. If in this case that is actually achieved, then the trick will presumably be done: either by distinguishing two quite different senses of ' socialism ', one for each of the two elements of the ostensible contradiction: or else by so qualifying both of these elements that they now refer to quite different aspects or parts of that party.

1.31 Because of the Hegelian or Hegelian-Marxist confusion involved in speaking of contradictions in things, and because salutary challenges to resolve seeming but not actual contradictions may be preposterously misconstrued as reasons for rating actual contradictions as in themselves good, contradiction sometimes wins an undeservedly favourable press. Similar confusions and misunderstandings often get an understandably bad one for logic.

1.32 The first of these misunderstandings hinges on a failure to distinguish two senses of the word ' logic '. One is primary. It is the sense in which the word has been employed in this chapter up till now. The other is secondary and derivative. This is the sense in which it is true to say that Aristotle, in the works grouped into what was later called the *Organon*, created Logic as an academic discipline. These two senses are most conveniently distinguished by printing the word with an initial capital whenever it is used in the second.

1.33 The general mistake here is that of expecting any study of that kind to be either necessary or sufficient to improve the practice to which it is directed. The musicologist does not through his musicology become a better executant. Nor is being a great performer immediately to be qualified as a musicologist. The particular point about Logic and logic was made in the late 1600s by John Locke in his *Essay concerning Human Understanding*: ' God has not been so sparing to men to make them barely two-legged creatures, and left it to Aristotle to make them rational. . . .' (IV (xvii) 4).

1.34 On the contrary: Logic, as the theoretical study of the forms and principles of argument, could only begin among and be pursued by people possessing a good practical capacity to separate valid from invalid arguments. In fact it grew up among the supremely argumentative Athenians. The present book, which is intended to help people to improve their thinking, is not an essay in theoretical Logic. It is instead an exercise in logical coaching. Such an exercise may be beneficial even though neither the coach nor the coached have or acquire any familiarity with the calculi of Logic. But it could not even begin, much less be beneficial, unless all concerned possessed at least some minimal competence in discerning soundness in argument. Without that you could not even understand the coaching.

1.35 The second and much more important reason why logic gets a bad press is that it is confused with various things which have nothing to do with it. Thus in an earlier period, when the British had—and perhaps with reason—more national pride, British leaders used to contrast illogical but practical British ways and British institutions with their always logical but often unworkable opposite numbers on the other side of the English Channel. Whatever else might be said for or against such contrasts there certainly is nothing logical, in our primary sense, about impractical practices or unworkable institutions. Nor, again in that sense, could there be anything to be proud of in being illogical. For that must involve failing to see what does or does not follow from what, and which propositions are or are not compatible. To be illogical, therefore, is to be stupid, or to be incoherent, or to be insufficiently concerned about truth, or all three together.

1.36 It was not for these faults that those proud leaders of former days were wont to commend themselves and their compatriots. What they saw, or thought they saw, was a contrast and a conflict between two opposite approaches to politics and to society. On the one side are those who like Plato want to start with a clean canvas, designing ideally orderly institutions in the abstract. On the other side are those like Aristotle who prefer to begin with whatever we have, seeing improvement as a matter of natural growth and development. The former are likely to long for utopia, and to have a penchant for wholesale operations. The latter do not expect anything to be perfect, and believe that whatever progress can be made has to be made piecemeal.

1.37 In the context of the great French Revolution of 1789 Edmund Burke represented the second approach, the one which with some reason is often thought to be typically English; while the Abbé Sieyès stood for the first, which is frequently with equal reason seen as characteristically French. Neither of these approaches is as such either logical or illogical, although the individual spokesmen of both may be. But when the Abbé Sieyès argues against legislative second chambers, his supporters are apt to applaud his famous apothegm as a fine specimen of Gallic logic, while his opponents upon exactly the same ground decry it. What he said was: that if the second chamber agrees with the first, it is superfluous, while if it disagrees with it, it is obnoxious.

1.38 It is not for us here to decide whether this is especially Gallic. We do, however, need to notice that it certainly is not especially logical. If someone accepted the two conditional propositions so dogmatically asserted, then it would, of course, be illogical for him to refuse to allow that any second chamber agreeing with the first must be superfluous, and that any second chamber disagreeing with the first must be obnoxious. Yet there is no intellectual or other merit in simply asserting these drastic propositions. By doing so you make, without supporting, the totalitarian assumption that all dissent on any point must be immediately and automatically overridden. This is a sort of assumption which no one has any business to make tacitly, and without argument. When such a brilliant refusal to examine the case for the opposition is presented as a model of logic, then there is every excuse to be suspicious. But it is not in this interpretation that we are labouring to make ourselves more logical.

1.39 Something has been said to show why contradiction ought to be unacceptable; and that logic is connected, albeit indirectly, with truth. It should now be less misleading to insist again upon the fundamental difference between questions about validity and questions about truth. To fix this in mind we need one or two dull, undistracting, naggingly unforgettable examples. So take as premises the two propositions: that *All philosophers are lifelong bachelors*; and that *King Henry VIII of England was a philosopher*. Both are false. But if they were true, then it would follow necessarily that *King Henry VIII of England was a lifelong bachelor*. For anyone asserting the two false premises, and yet denying the conclusion, would certainly be committing a self-contradiction. So now, since few people have less claim to have been lifelong bachelors than that six times married English king, we have an example of a valid argument from false premises to a false conclusion.

1.40 Suppose next, that, following a strictly mid-Atlantic course, we substitute for the original second premise: that *President Buchanan was a philosopher*. Then again, both premises will be false. The conclusion will be, that *President Buchanan was a lifelong bachelor*: and, applying the same test as before, it is obvious that this is derived by a valid deductive argument. But this time the conclusion is, I am told, true. So now we have an example of a valid argument from false premises to a true conclusion.

1.41 Someone says: that *All Communists claim to repudiate racial discrimination*; and that *Dr Angela Davies claims to repudiate racial discrimination*. If we accept these two premises, are we required and entitled to infer that *Dr Angela Davies is a Communist*? No, we are not. Certainly the con-

clusion was, at the time of writing, true. With the expression
' claim to ' carefully inserted, to allow for such scandalous
realities as the anti-semitism tolerated or inspired by the
Soviet Communist Party, the premises are no doubt also
true. Yet the argument is still invalid. To make it valid,
the first premise would have to be changed to read: not
' All Communists '; but ' All and only Communists '. In
fact most of those who not merely claim to repudiate all
(repeat all) racial discrimination, but genuinely do, are not
Communists. So the example offered, without the necessary
amendment, constitutes an example of an invalid argument
to a conclusion which happens nevertheless to be true.

1.42 Suppose that someone has difficulty, as many do, in
appreciating that such an argument must be invalid. Then
one natural and right response is to summon up parallels to
enable them to see that neither this nor any other argument
of the same form can possibly pass. You might as well
argue, we might say, that: given that *All swans are white*
(which, as Australians know, they are not), and given that
Governor Wallace is white (which, in a sense, he is); then, it
follows necessarily, *Governor Wallace is a swan* (which will
scarcely do). Or again, you might as well argue: given that
All Communists claim to repudiate racial discrimination, and given
that *Dr Martin Luther King claimed to repudiate racial discrimination*; then, it follows necessarily, *Dr Martin Luther
King was a Communist*.

1.43 What we are doing when we produce such parallels
is trying to bring out the invalidity of all arguments of one
particular form; the form, that is, and whatever it is, which
is shared by both the original specimen and all the genuine
parallels which could be deployed. The main practical
reasons why parallels have to be summoned is that people
get put off by what they know or believe about particular
propositions in particular arguments. Because we know or
believe that the proposed conclusion is true, we become less
alert to the possible weakness of the inference by which it is
supposedly derived. If, therefore, we want to test someone's

critical acumen, it is best to attend to their responses to the discourse of those with whom they themselves are most inclined to sympathize.

1.44 Suppose, for instance, I am like most left-thinking people a socialist. Then try me to see whether I am embarrassed to hear someone urging—as one often may—that such and such an industry is dominated by three or four big monopolies; and hence that it should be nationalized forthwith. Whatever the merits of the conclusion, this is a quite lamentable argument. To start with, and it is an egregiously bad start, the premise is self-contradictory: if there are three or four big firms competing, then no one of them can have a monopoly, by definition. Furthermore, if we actually are confronted with the evils of monopoly, it is scarcely sensible to propose as a remedy the immediate creation of a (state) monopoly.

1.45 A measure of symbolization may now be practically useful. What the fallacious arguments of paragraphs 1.41 and 1.42 have in common is the following form. Given that *All so-and-so's are such-and-such*, and given that *That is a such-and-such*; then, it follows necessarily, *That is a so-and-so*. It is a very short, and a space-saving, further step to replace ' so-and-so ', ' such-and-such ', and ' That ' by letters. If you are going to do this, now and or later, then it is also a good thing to introduce the further notational refinement of distinguishing the subjects (the so-and-so's) from the characteristics attributed to these subjects (that of being such-and-such), by employing capital letters from our ordinary Latin alphabet for the former, and by calling on lower case Greek letters to symbolize the latter. Thus: If *All A's are* ϕ (pronounced phi), and if *That is* ϕ, then, it follows necessarily, *That is an A*; which, of course, it does not.

1.46 So much for the key notion of the form of an argument. Here and elsewhere all the particular specimens of any general class may be described as the several tokens of that same single type. The particular type or form of argument of which we have just been considering some tokens is fallacious. It has an unfortunately unmemorable traditional name. The name is ' The Fallacy of the Undistributed Middle '.

1.47 In the heyday of Senator Joseph McCarthy and of the House Committee on Un-American Activities some favoured the nickname ' The Un-American Fallacy '. This was a backhanded tribute to that Senator and to those members of his committee who were inclined to deduce that a person must be a Communist from the evidence that he possessed some characteristic perhaps shared by all Communists, but certainly not in fact peculiar to them. This particular nickname is long since obsolete. Yet we still need to take the point made by the pessimistic German philosopher Arthur Schopenhauer, though discounting his false and nasty suggestion that every defect from logical perfection is studied and designing: ' It would be a good thing if every trick could receive some short and obviously appropriate name, so that when a man used this or that particular trick, he could be at once reproached for it ' (Schopenhauer, p. 18).

1.48 In the present context the word ' fallacy ' does not refer to just any intellectual error. It is confined to one particular sort of such errors, that of mistaking an invalid for a valid argument. This needs to be emphasized, since there is a common usage in which any false proposition may be described as a fallacy. Thus in 1973, the first year after joining the European Economic Community (EEC—the Common Market), Britain suffered a huge and horrid increase in food prices. Probably a majority of the British people believed that this was largely, if not entirely, due to the enforcement of the Common Agricultural Policy (CAP). The minority, who held that belief to be mistaken, could and did say, in accordance with this common usage, that it was a fallacy.

1.49 If it were only a matter of what is acceptable to dictionary makers as established, and hence correct,

English usage, then the pro-EEC minority could have rested their case for employing the word 'fallacy', rather than the equally wounding 'misconception', upon the undoubted propriety in these dictionary terms of the label 'The Pathetic Fallacy'. This label refers to the mistake of believing that something, which is not a person at all, nevertheless has plans and purposes and feelings such as only a person could have. But, in our stricter sense of the word 'fallacy', neither this nor the putative misconception about the cause of the 1973 rise in food prices is a fallacy. The fallacy involved, if fallacy there was, must have been, not in the conclusion, but in the supporting argument.

1.50 In the EEC case many of those who accepted the conclusion uncritically certainly were committing a fallacy; which is one reason for choosing that as the example here. The authentic fallacy was that of arguing that, simply because that occurred after this, therefore, that must have occurred on account of this. Clearly such an argument will not do. For the assertion that the beginning of a steep rise in food prices occurred after Britain's admission is compatible with the contention that this was purely coincidental, and that the actual cause of that rise lay elsewhere. This fallacy has been known traditionally—translating the key phrase into the pontificating obscurity of a learned language—as that of *post hoc ergo propter hoc*. Until and unless someone is able to suggest a better English alternative, let us call it The Whatever-follows-must-be-the-consequence Fallacy.

1.51 The prime reasons for insisting upon the stricter usage of the term 'fallacy' are efficiency and economy. We have in our rich language other words which can be used for just any kind of mistake or misconception. For a start there are the words 'mistake' and 'misconception'. If we oafishly misemploy our verbal chisels as verbal screwdrivers, we thereby unfit them for the job to which they are best suited. So what do we use for a chisel when a chisel is what we need?

1.52 Compare another topical example, far removed from our immediate interests. Those who have enjoyed such classic gangster movies as *The Roaring Twenties* will remember that the word ' hijack 'was first introduced to refer to the forceful seizure of what was already stolen or in some other way contraband. There is, surely, nothing to be said for the current abusage, which makes ' hijacking ' a superfluous synonym for the redolent and exciting 'piracy ', and thus leaves us without any handy single word for the original Prohibition phenomenon. What term are these sloppy and cinematically illiterate journalists going to use suppose a stolen car driven by the air pirates of Black September is, in the strict old gangster sense, hijacked by the militant henchmen of Rabbi Meir Kehane, perhaps on their way to blow up an Aeroflot Concordski? That will be the day!

1.53 It is for similar reasons that I have been following, and shall continue to follow, stricter usages of many other everyday terms. Such stricter usages are required even in making and maintaining the fundamental distinction between questions about truth or falsity and questions about validity or invalidity. Nor is there any call to go slumming in order to unearth examples of what we have got to avoid. In his *Discourse on the Method* the Father of Modern Philosophy formulates his proposed doubt-proof, rock-bottom certainty as an argument: ' I think, therefore I am.' Yet he still affirms that this is something which he ' clearly and distinctly conceives to be true ' (Part IV). Allowance must of course be made in Descartes' case for the fact that Descartes was writing in the earlier 1600s. But that is a reason why we have to do better.

1.54 Some other instances have already been noticed. Care is also always required about knowledge and refutation. To say that someone knows something is to say more than that he claims to know it, or that he believes it most strongly. It is to say also, both that it is true, and that he is in a position to know. So neither the sincerity of his conviction nor the ingenuousness of his utterance guarantees that he knew. That is why the sarcastic tone enters our voices, or why we write the key word between sneer quotes, when a man who has claimed to know turns out to have been wrong: ' He " knew " what was going to win the 14.30, but he " knew " wrong.' To say that Wilson refuted Heath is to say more than that Wilson asserted that Heath was in error. It is to say also, both that Heath in fact was in error, and that Wilson proved that this was so.

1.55 If you do not want to say as much as that, then you should take the trouble to be non-committal. You ought to say only: in the one case, that he believed, and claimed to know; and, in the other case, that Wilson denied what Heath had maintained, and considered that he had refuted it. The same desire to husband resources of vocabulary, to preserve vital distinctions, should make us stingy in our application of the term ' prejudice '. Often it is treated as roughly equivalent to ' opinion ' or ' conviction ', albeit with powerful pejorative overtones. In this all too common abusage I have my opinions and my convictions, but you and he merely have prejudices—so called for no better reason than that they are yours or his and not mine.

1.56 The word ' prejudice ' becomes a valuable extra item in the vocabulary of anyone striving to be more rational only when, and in so far as, it is employed scrupulously to pick out just those beliefs—whether right or wrong —which are either formed prior to proper consideration of the evidence or else maintained in defiance of it. It is obscurantist and demoralizing to apply the word in order to abuse other people's opinions, or even all strong convictions simply as such. The Judge who instructs the jury to consider carefully, and without prejudice, all and only the materials actually presented in court, is not asking them to refuse to bring in a decisive verdict. Nor is there anything whatever wrong with anyone's opinions or with strong convictions, as such. What is obnoxious, and what merits all the abuse in the stockpile, is the wilful maintaining of preconceptions against the weight of the evidence. But to do that is not an always incurable feature of the human condition. Nor is it the exclusive prerogative of other people.

2 If/then and All/none

2.1 Some examples in Chapter One were, for good reasons given, trifling and even frivolous. Yet nothing could be more wrong than to carry away the impressions that logic itself is trifling, or that arguments cannot refer to matters of life and death. In his last letter to his wife one of the doomed soldiers of the German Sixth Army outside Stalingrad wrote: ' If there is a God, you wrote to me in your last letter, then he will bring you back to me soon and healthy. . . . But, dearest, if your words are weighed now . . . you will have to make a difficult and great decision. . . .' (Schneider and Gullans, p. 59).

2.2 The soldier was right. For if from any hypothesis, in this case that of the existence of (a certain sort of) God, you can validly deduce some consequence, in this case that a particular soldier will return ' soon and healthy ', and if the consequence thus validly deduced is false; then it follows necessarily that the hypothesis itself must be false too. So the wife, when she hears the news of the death of her husband, will indeed ' have to make a difficult and great decision '. She will have to decide whether simply to abandon the hypothesis, or whether instead so to reinterpret it that it does not entail the kind of consequences which she had thought to be validly deduced therefrom.

2.3 In the one case she will argue: Since *If there were a God, then He would ensure that my husband would return safe from the wars*, and since *My husband will not return safe from the wars*, then, it follows necessarily, *There is no God*. In the other case she will, as we all must, accept that this last argument is valid. But she will now deny the truth of the conditional proposition which functions as the first premise of that argument: *If there were (is) a God, then He would (will) ensure that my husband would (will) return safe from the wars*. What will be called continuing to believe in God will, therefore, involve for her a reinterpretation of the word ' God '. Previously she must have been construing it in such a way that it followed, from a statement that *There is a God*, combined no doubt with the assertion of some other premises here unstated, that *He will ensure my husband's safe return*. But now she has come to interpret the same word ' God ' in such a way that this inference is invalid: there is, that is to say, no contradiction involved in at one and the same time, both asserting these premises (properly understood), and denying the in fact false conclusion.

2.4 Arguments of the same form as that of the soldier at Stalingrad are employed wherever any hypothesis is tested, both in science proper and everywhere else. Sherlock Holmes, for instance, might argue: since *If the thief entered this way, then there would be footprints in the flowerbed*, and, since *There are no footprints*; therefore, I deduce, *The thief did not enter this way*. The form of the argument is, and it is a valid form: If *This is so* then *That is so*; but *That is not so*; and, therefore, *This is not so* either. Since *This is so* and *That is so* are propositions, and not subjects or predicates, it is sensible, if you want to introduce symbolism, to make your symbolism appropriately different from that suggested earlier for subjects and predicates (1.45). The established convention is to use the lower case Roman letters, p, q, and r. Since *This is not so* and *That is not so* are simply the denials or negations of, respectively, *This is so* and *That is so*, the further convention is to symbolize the negations of p, q, and r as, respectively, ~p, ~q, and ~r (read as not p, not q, and not r). Putting this new symbolism to work at once, we can now express the present valid form of argument as follows: *If p then q, but ~ q, therefore ~p.*

2.5 Let us now distinguish this important and valid form from two other forms which are invalid. Let the two conditionals again be granted: *If there is a God, then the soldier will return*; and *If the thief entered this way, then there will be footprints in the flowerbed.* Now suppose: that *The soldier has returned*; and that *There are footprints in the flowerbed.* It will not do to infer from these pairs of premises the conclusions, that *There is a God*, and that *The thief entered this way.* In such very simple cases, in which none of our public or private concerns are engaged, it is easy to see why it will not do. The crux is that the premises so far provided do not preclude the possibility of some alternative explanation. Even in a Godless world some soldiers would presumably return. So why not this one? Certainly, had the thief entered this way, then he would have left his footprints in the flowerbed. But nothing has so far been said to rule out the possibility that any footprints found there might have been made by someone other than the thief.

2.6 The second of the two invalid moves approaches from the opposite direction. Let the two conditionals be granted yet again: *If there is a God then the soldier will return*; and *If the thief entered this way, then there will be footprints in the flowerbed.* But now suppose: that *There is no God*; and that *The thief did not enter this way.* It will not do to infer from these fresh pairs of premises the conclusions, that *The soldier will not return*, and that *There will be no footprints in the flowerbed.* The crux this time is that such conditionals say nothing about the situation on the alternative assumption that their antecedents are not true. It may well be true that, if you become a schoolteacher, then you will be short of money. But from this we are not entitled to infer that you will not be short of money if you do anything else. The chances are you will!

2.7 Four possible moves can be made with such conditional propositions, two valid and two invalid. Some readers may find it helpful to run through the quadruplet systematically. But if for you the emphasis in this sort of foursome reel is too much on the word 'reel', then skip straight on to the next paragraph but one.

2.8 The valid move of arguing *If p then q, but ~ q, therefore ~ p* involves denying the consequent (the then bit) of the conditional in order to prove the antecedent (the if bit): hence, for short, Denying the Consequent. The first invalid move, illustrated in paragraph 2.5, consists in arguing *If p then q, but q, therefore p*. This involves affirming the consequent in baseless hope of thereby proving the antecedent: hence, for short, Affirming the Consequent. A second valid move is too obvious, and too obviously sound, to require illustration. This is to argue: *If p then q, p, therefore q*. In this we affirm the antecedent in order to prove the consequent: hence, for short, Affirming the Antecedent. The second invalid move completes the ring of changes. This invalid move, illustrated in paragraph 2.6, is to argue: *If p then q, ~ p, therefore ~ q*. Here we deny the antecedent in the baseless hope of thereby disproving the consequent: hence, for short, Denying the Antecedent.

2.9 Important and fundamental questions about scientific method turn on elementary logical points expounded in the previous eight paragraphs. The heart of the matter lies in an asymmetry with regard to verification and falsification. Hypotheses are tested by deducing consequences which would follow if the hypothesis were true. When a hypothesis is tested, and even one of its consequences is found in fact not to obtain; then that hypothesis, at least as originally formulated, is decisively demonstrated to have been false. Thus the original hypothesis entertained by the soldier's wife tested in the inferno of Stalingrad, was decisively disproved by her husband's death there. Holmes's hypothesis that the thief came thisaway was similarly falsified—shown, that is, to be false—when no footprints were found in the flowerbed. But when a hypothesis is tested, and one of its consequences is in fact found to obtain; then it is still not thereby demonstrated that that hypothesis must be true. Suppose her husband had returned safe and sound, this would not have proved the existence of her God. Now would the discovery of footprints in the flowerbed have ruled out the possibility that these were made by someone else, and that the thief himself used a different route.

2.10 This asymmetry with respect to verification and falsification also characterizes such open universal propositions as *All swans are white* or *All living creatures are mortal.* These are called universal because they refer to all the members of a class; and open because that class is one the members of which cannot be exhaustively listed. A single genuine counter-example is sufficient to disprove any such proposition. But, no matter how many confirming examples you produce, it is impossible to prove it with any correspondingly decisive finality. It was this logical observation which led Francis Bacon to make his often quoted remark that ' the force of the negative instance is greater '. (Bacon, by the way, fell from the high dignity of Lord Chancellor of England in the early 1600s as a result of what was known as the affair of the Water Gate. Apparently this Water Gate, as I am sure that the British Travel and Holidays Association would wish me to mention, can still be seen today in the City of London.)

2.11 It is the observation of these logical asymmetries which serves as one foundation of Sir Karl Popper's enormously influential and salutary philosophy of science. No open universal proposition can ever be confirmed beyond all possibility of future correction, although some can be decisively and finally disproved. Yet the crucial elements in every conceivable scientific theory, and any acceptable candidates for the status of law of nature, must be propositions of this kind. So Popper draws the exciting moral that science is and can only be a matter of endless striving and of endless inquiry. No law or theory can ever stand altogether beyond the possibility of revision. The best we have, or ever could have, can, and could only be, the best so far (Magee, and Popper (2)).

2.12 We are bound to be distressed—if we are ourselves

big enough to revere greatness—to discover that Popper has himself contributed an apt and interesting example of the fallacy of Denying the Antecedent. Yet the right lesson to learn from his embarrassing lapse is utterly Popperian: the need for unresting critical alertness—perhaps especially in our attention to those with whom we most agree and whom we most admire.

2.13 The lapse occurs when Popper is reconsidering that medieval favourite *All men are mortal*. He remarks that this ' mortal ' ought to be construed as ' liable to die '. This is certainly true. It was because they were in fact construing it as (naturally) ' liable to die ', that those who believed in the (miraculous) translation both of the Prophet Elisha and of Mary the mother of Jesus rightly refused to accept these as counter-examples. The claim that these two exceptional people were by a miraculous overriding of the laws of nature taken straight up into heaven does not deny but takes it for granted that they too, like all other humans, were naturally liable to death. But for those miracles they too would in due course have died ordinary unprivileged natural deaths. Popper next proceeds to report that it was ' part of Aristotle's theory that every generated creature is bound to decay and die. . . . But this theory was refuted by the discovery that bacteria are not bound to die, since multiplication by fission is not death . . .' (Popper (3), p. 10). Certainly from the hypothesis that *All living things are bound to decay and die* we can deduce: first, *All bacteria are bound to decay and die*; second, *All humans are bound to decay and die*; and so on indefinitely. And certainly the falsity of the first of these consequents is quite sufficient to show the falsity of that antecedent. But it is fallacious to argue, from the falsity of that antecedent, to the falsity of any of those consequents. Yet that is precisely how Popper does argue here: ' The second was also refuted, though not as obviously.'

2.14 The facts, that *All men are mortal* is indeed a consequence of *All living things (including men) are mortal*, and that because of the splitting bacteria the latter is now known to be false, are not sufficient reasons for saying that the former must be false also. Suppose they were. Then to refute any theory would be simultaneously to refute all the propositions deducible from that theory, and hence all the propositions which it was capable of explaining. This would carry a curious and a catastrophic consequence. Any established theory which is refuted, is refuted through its failure to explain recalcitrant facts. And it would surely be unusual, and certainly scandalous, if a theory were to have become established notwithstanding that it could not explain any even of the facts already known in its heyday. On the present supposition, however, we should by refuting any established theory be left with nothing for any new theory to explain except the fresh facts deployed in that very refutation. Which is absurd.

2.15 The four moves listed systematically in paragraph 2.8 can be further illuminated with the aid of distinctions between necessary and sufficient conditions, distinctions far and away more valuable than any of the eminently forgettable technicalities of that paragraph. The distinction which I am about to explain is not exactly that put to work in paragraph 2.19; and is different again from that of paragraphs 2.23-2.26. The immediate object is to give the general hang of such distinctions by displaying a particularly fine specimen. Anyone in no mind to mind his p's and q's should skip on at once to those later paragraphs.

2.16 When, given the truth of one proposition p, the truth of another proposition q follows necessarily, then the truth of p is, by definition, a logically sufficient condition of the truth of q. That Bertie is an Englishman is thus a logically sufficient condition of the truth that Bertie is a man. When some proposition r necessarily cannot be true unless some other proposition s is also true, then the truth of s is, again by definition, a logically necessary condition of the truth of r. That Cynthia has been married is a logically necessary condition of the truth of the proposition that

Cynthia is divorced. You just cannot even begin to get divorced unless you have first got married.

2.17 Obviously to say that p is a logically sufficient condition of q is not to say: either that p is the only sufficient condition; or that it is a necessary condition. Since Frenchmen are also men, that Bertie is a Frenchman would be another, alternative, logically sufficient condition of the truth that Bertie is a man. For the same good and sufficient reason it is not a necessary condition of the truth that Bertie is a man that Bertie is an Englishman.

2.18 Again, to say that s is a logically necessary condition of r is not to say: either that it is the only logically necessary condition; or that it is a logically sufficient condition. Since it is not the case—not even in such pace-setting countries as Sweden and the USA—that everyone who gets married forthwith becomes divorced, it may well be true that Cynthia has married and has not been divorced. And even if everyone getting married did in fact proceed at once to divorce, still to say that someone has married would be to say less than to say that they are now divorced. So that she got married is a logically necessary but not the only logically necessary nor the logically sufficient condition of her being, as of now, divorced. (We have all, surely, known marriages at which it would have been fair, though unseemly, to comment: ' Well, I suppose marriage is a logically necessary condition of divorce '.)

2.19 Putting a distinction between necessary and sufficient conditions immediately to work, we can say that the two fallacies represent misunderstandings of what it is for one proposition to be any sort of sufficient condition of another. For to say *If p then q* just is to say that p is some sort of sufficient condition of q. The second invalid move, to argue *If p then q, but ~p, therefore ~q* confounds this initial statement that p is a sufficient condition of q with the claim that p is a necessary condition of q: were p indeed a necessary condition of q then we could infer ~q from ~p. The first invalid move, to argue *If p then q, but q, therefore p,*

confounds the same initial statement that p is a sufficient
condition of q with the claim that p is the only sufficient and
necessary condition of q; were p indeed the only sufficient
and necessary condition of q then we could infer p from q.
Or, in other words, this second fallacy consists in the
confusion of *If p then q* with *If and only if p then q*. Since
English – like, so far as I know, all other natural languages –
lacks a single word for if and only if, Logicians have invented
the artificial and seemingly misspelt vocable ' iff '. In these
terms this second fallacy of Denying the Antecedent can be
characterized as the mistake of deducing from *If p then q*
what would indeed follow from *Iff p then q*.

2.20 If things seem to have been moving either too fast or
too abstractly in the five previous paragraphs, and especially
if you think of yourself as a student rather than as a general
reader, then you would be well advised to settle down with a
pencil and paper to run through the various logical points
about conditionals treated in all the first sixteen paragraphs
of the present chapter, in your own way and at your own
pace. Half the point of having, and using, the simple
symbolisms introduced in paragraph 2.4, and those introduced earlier in paragraph 1.45, is to make this sort of
checking easier and quicker. (The other half is to escape
the often misleading distractions of individual interests in
and beliefs about some of the particular propositions which
may be serving as premises or conclusions in some doubt-
fully valid argument.) Certainly no one can reasonably
expect to acquire any facility even with this very modest
minimum of symbolism without undertaking a little private
practice of this sort. In general, any argument which is
either difficult or contentious should be examined closely in
writing. It is altogether too easy—even when, which is not
always the case, the intentions of all concerned are impec-
cable—for the swiftness and the fleetingness of the spoken
word to deceive the mind. And, just because what is not
recorded is not recorded, everyone's ground may shift
without anyone's noticing.

2.21 Suppose someone says, as someone often does: *If I (Jones) do not get this, then there is no justice in this world.* And suppose, as sometimes happens, he does in fact get what he wants: *Jones gets this.* Are we entitled from these two premises alone to infer the conclusion which someone else is sure now to point to Jones: *There is some justice in the world?* Certainly this conclusion is true, and perhaps justice did indeed demand that Jones got this. But this argument as an argument—and that is what we are talking about—is invalid. It is an instance of (attempting to disprove the consequent by) Denying the Antecedent.

2.22 There are three reasons why it is perhaps not immediately obvious that the particular argument about Jones is fallacious: first, the proposed conclusion is in fact true; second, we get hazed by the negatives in both antecedent and consequent; and, third, the fact that justice was done on this occasion—assuming it was—is by itself and without recourse to any other argument a sufficient reason for saying that *There is some justice in the world.* We remove all three difficulties if we get that pencil and paper, and write down: p = *I (Jones) do not get this*; and q = *There is no justice in this world.* Given these interpretations of p and q we can now symbolize the argument: *If p then q, but ~p, therefore ~q.* And this is manifestly a form of argument the invalidity of which we came fully to appreciate earlier; especially in paragraphs 2.6, 2.8, 2.12-14, and 2.16. Of course, we have also to appreciate that with these interpretations of p and q the negations ~p and ~q become, respectively, *Jones gets this* and *There is some justice in the world*: the rule is that double negation, not not, cancels out, leaving no negation at all.

2.23 There is another distinction between necessary and sufficient conditions which is of great practical importance. When we say that p is a logically sufficient condition of q, then we are saying that we could not assert p and deny q without contradicting ourselves. But when such and such is said to be not the logically but the causally sufficient condition of this or that, then it is being said that, in the world as it actually is, with the laws of nature what they actually are, such and such could not in fact be produced without producing this or that. When such and such is said to be not the logically but the causally necessary condition of this or that, then it is being said that, in the world as it actually is, with the laws of nature as they actually are, this or that could not in fact be produced without first producing or having such and such.

2.24 Suppose that heavy smoking is, in certain circumstances, and for people of a certain constitution, a causally sufficient condition of eventually contracting lung cancer. Then this means that, in those circumstances, and for those people, heavy smoking in fact guarantees eventual lung cancer. We need not suppose, but can know, that it is not in general a causally necessary condition. For there certainly are some lung cancer victims who have never smoked at all. It is, however, not contradictory to suggest that what happens to be in our world either causally sufficient or causally necessary conditions, would in a different universe be neither. Or, putting exactly the same point in different words, there is nothing inconceivable about the idea of a universe governed by quite different laws from those which seemingly govern the actual universe. Or, in yet other words again, the occurrence of an event which must be, according to these actual laws, physically impossible is by no means by that token logically impossible (1.24 and 1.25). Indeed there seems to be no contradiction in the suggestion of a universe not governed by any laws at all—a point made, to the great scandal of the philosophical public, by David Hume in *A Treatise of Human Nature* (I (iii) 3).

2.25 To put this further distinction at once to work, consider one argument offered by a pair of contributors to a widely circulated volume published by Penguin Education. They are attacking the proposition that heredity matters: ' The model, and especially Galton's version of it, of course denies the possibility of change . . . if 80 per cent of adult performance is directly dependent upon genetic inheritance, how have the styles of our lives and the patterns of our thinking changed to the extent that they have? ' (Richardson and Spears, p. 74).

2.26 This objection, which its proud authors obviously regarded as knock-down decisive, is simply irrelevant. For what their opponents were maintaining is that some appropriate level of ability is a causally necessary condition of any kind of achievement; and that what abilities are available to each individual is largely—to the extent of about 80 per cent—determined by that individual's heredity. To put it somewhat coarsely, what these opponents hold is: that you cannot make silk purses out of sows' ears; not that silk purses are born not made. The objection offered would be decisive against the second contention, the sufficient condition thesis. But that thesis no one has ever maintained —certainly no one of the scientific calibre of Darwin's cousin, Sir Francis Galton F.R.S. When the actual contention is one about putative necessary conditions, which it is, the attempt to overthrow it without benefit of any grasp of the crucial distinction cannot do anything but reveal the poverty of the intellectual equipment of the assailants.

2.27 There is a further and wider lesson to be drawn from this spectacle. It is that where we are ourselves committed to an extreme position, we are apt to assume, or to pretend, that everyone else is committed to some equally extreme position; which is rarely true. This is a special case of the general temptation to attribute to our opponents positions even more absurd and more easily refutable than those which they do in fact actually maintain. We much too easily persuade ourselves that ' those people ', in the other party, hold that such and such is a panacea, which will by itself produce all good things. But perhaps what they are actually claiming is, more modestly: not that it is the sufficient condition; but that it is a, or at most the, necessary condition. Enemies of socialism, for instance, are thus too apt to attribute to all socialists the conviction that the nationalization of all the means of production, distribution, and exchange is the sufficient condition of the building of Blake's new ' Jerusalem in England's green and pleasant land. '

2.28 From the opposite side of the House many will join with the Labour veteran Mr Emmanuel (now Lord) Shinwell in urging: ' there can be no valid objection by the Tory Party against the principle of nationalization of steel or any other industry unless, should they form a government, they would transfer the existing nationalized industries . . . to private ownership ' (*The Times* 11/iv/66). But by agreeing that, in certain circumstances, some particular industry or industries ought to be, or to remain, nationalized you do not become committed to agreeing that some other industry should be, or remain, nationalized; much less that, immediately or ultimately, all industry should be nationalized ' on principle '. What, of course, you cannot maintain consistent with any such concession is that no industry ought to be or to remain nationalized, in any circumstances, ever. What Lord Shinwell thinks of as ' the principle of nationalization ' is, presumably, the former: the socialist demand for total public ownership of (to quote the words of the original Clause IV of the Constitution of the Labour Party) ' all the means of production, distribution and exchange'. But to reject this socialist principle is not at all to accept the extreme contrary position: that no industry ought to be or to remain nationalized, in any circumstances, ever. And though there are many socialists devoted to the former, few if any of their actual opponents in fact maintain the latter.

2.29 Again, returning to the horrid example of paragraphs 2.25 and 2.26, those inclined to hold that everything is environmentally determined, become for that very reason tempted either to assume or to pretend that anyone who disagrees with them must hold that nothing is environ-

43

mentally determined. Certainly, if this polarization assumption were universally true, it would make it easier for all spokesmen of extreme positions to dispose of the thus correspondingly extreme positions of their opponents. Hence, of course, the strength of the temptation. But the controversial world is not in fact so conveniently arranged, with this sort of ideal balance. In the present case, for instance, while there never have been—and for very good reason—any extreme hereditarians maintaining that our genetic constitution is the sufficient condition of all that we are and can be; there certainly are at this time many environmentalist ultras. These do want to hold that everything about us—or as near as makes precious little matter everything—is determined by our environments.

2.30 These points about the temptations of extremism neither depend upon nor do they warrant any general contention that we ought to strive for moderation in the content of all our opinions, or that truth is always in the middle. Suppose that it had been—say—President Kennedy and not, as it was, Senator Goldwater who said: '. . . extremism in the defence of liberty is no vice . . . moderation in the pursuit of justice is no virtue' (Acceptance speech at the Republican Convention, 16/vii/64). Then all America's liberals—and not they alone—would have endorsed these splendid words. Such hypothetical endorsers would have been right. We should be right to concur anyway. Again, it actually was Abraham Lincoln himself who warned: ' Let us not be diverted by more of these sophistical contrivances wherewith we are so industriously plied and belaboured—contrivances such as groping for some middle ground between the right and the wrong' (Speech at the Cooper Union, New York City, 27/ii/1860).

2.31 The alleged truism that truth is always in the middle is not merely false but demonstrably false. If the truth really were, as this silly and unprincipled principle requires, half way between A and Z, for all values of A and Z; then it must also be, according to the requirements of the same principle, half way between the half way point between A and Z and Z; and so on, indefinitely. This conclusion is incoherent and absurd. The same must, therefore, apply to the principle from which it is thus validly deduced: The Truth-is-always-in-the-middle Damper.

2.32 The abstract theoretical demonstration has concrete practical relevance. For there are in fact a great many people who adopt as one of their main guiding principles, that they should, with regard to all controverted issues of belief, strive to position themselves equidistant between whatever they see as the most extreme standpoints currently adopted. Not only does this, for reasons just now stated or suggested, guarantee commitments to error of all kinds. It also—paradoxically—exposes its protagonists to manipulation from those very extremes which it professes to eschew. For persuasive operators, discerning this common affection for whatever can be presented as the middle ground can, and do, find ways of shifting the apparent centre in whatever direction they desire. Weakminded middlemen must as such construe the fact that some are pushing in the A direction, while there happens to be no equal and opposite group pushing in the Z direction, as a sufficient reason why they should themselves—at, of course whatever distance is required to avoid extremism—follow in the A train. This manipulation of the middlemen is the analogue in the sphere of belief and persuasion of a universally familiar and too often successful bargaining strategy, the strategy whereby one side makes inordinate demands in order that any splittings of the difference fifty-fifty shall yield a result going further its way than the other.

2.33 Whatever may be the case in bargaining situations, where the object just is to get more or less of whatever it may be, it is quite different where the aim is—or ought to be —truth. When that is the object, then considerations of moderation or extremism are as such neither here nor there. They can then find purchase in some particular context only in so far as we happen to have some substantial independent reason to believe that in that context a moderate thesis, or an extreme one, is more likely to be true.

2.34 There are, as we have just been noticing (2.30-2.32), insidious dangers and temptations in what presents itself, or is presented, as the middle ground. The temptation of the extremist, on the other hand, is ' to assume, or to pretend, that everyone else is committed to some equally extreme position ' (2.27). The particular extremists who occasioned this comment held, or at least hankered to hold (2.25-2.26), that: (1) *All human differences are determined by the environment.* To disagree with this it is sufficient to assert no more than that: (2) *Not all human differences are determined by the environment.* The first of these two propositions is a universal proposition; the second is its contradictory; and, as we have seen (2.10), any universal proposition can be decisively falsified, and its contradictory equally decisively verified, by the production of even one single genuine counter example. Yet our extremists, perhaps on the basis of an invalid argument, and for certain mistakenly, assumed that their opponents are committed to the correspondingly extreme and diametrically opposite thesis: (3) *No human differences are determined by the environment.*

2.35 This third proposition, which Logicians somewhat misleadingly dub the contrary of the first, is, like that first itself, but unlike the second, universal. It too, therefore, is exposed to definitive falsification by the production of a single genuine counter example. So the protagonist of the first is bound to see his task as easier if he believes, or pretends, that he has to refute only the third and not the second. Nevertheless it is, as must by now be obvious, fallacious to argue that, because someone denies the first, and hence is committed to the second, therefore he must in consistency assert the third. The contradictory is not the contrary.

3 Evasion and Falsification

3.1 Imagine some Scottish chauvinist settled down one Sunday morning with his customary copy of *The News of the World*. He reads the story under the headline, ' Sidcup Sex Maniac Strikes Again '. Our reader is, as he confidently expected, agreeably shocked: ' No Scot would do such a thing! ' Yet the very next Sunday he finds in that same favourite source a report of the even more scandalous on-goings of Mr Angus MacSporran in Aberdeen. This clearly constitutes a counter example, which definitively falsifies the universal proposition originally put forward. (' Falsifies ' here is, of course, simply the opposite of ' verifies '; and it therefore means ' shows to be false '.) Allowing that this is indeed such a counter example, he ought to withdraw; retreating perhaps to a rather weaker claim about most or some. But even an imaginary Scot is, like the rest of us, human; and we none of us always do what we ought to do. So what in fact he says is: ' No true Scotsman would do such a thing! '

3.2 An equally simple, but actual, example of this No-true-Scotsman Move was provided by the Black Power leader Stokely Carmichael during a visit to London some years ago. He was arguing the thesis that the world is now divided between exploiting white men and exploited coloured people: ' What about Castro? ' asked one member of his audience, ' What about Che Guevara? ' ' I don't ', retorted Mr Carmichael, ' consider them white ' (Reported by Peterborough in the *Daily Telegraph*, 20/vii/67).

3.3 In these two textbook examples it is immediately obvious what is going on, and wrong. A bold, indeed reckless, claim about all those who happen to be members of a certain category is being surreptitiously replaced by an utterance which is, in effect, made true by an arbitrary redefinition. If anyone who satisfies all the ordinary requirements for being accounted a Scot behaves like the Sidcup sex maniac, then our chauvinist will take that fact as by itself sufficient reason to disqualify him from rating as a true Scot: true Scots are, by definition, not sex maniacs. If anyone who would conventionally count as a Caucasian commends himself to Mr Carmichael, then that fact alone becomes for Mr Carmichael decisive against considering that exceptional Caucasian to be truly white. If all examples of the No-true-Scotsman Move were as simple and straightforward as these two, then there might perhaps have been little need to introduce such a label. But they are not.

3.4 The essence of the move consists in sliding between two radically different interpretations of the same or very similar forms of words. In one, in this case the original interpretation, what is asserted is synthetic and contingent. In the other, in this case the later and highly factitious interpretation, what we have is a made to measure necessary truth. This fundamental distinction can be brought out, and fixed securely in mind, with the help of Shakespeare. Questioned about what the ghost said Hamlet replies (*Hamlet* I (v)):—

> There's ne'er a villain dwelling in all Denmark
> But he's an arrant knave.

His friend Horatio responds:—

> There needs no ghost, my lord, come from the grave
> To tell us this.

3.5 There needs no ghost because Hamlet's proposition is analytic and logically necessary. What makes it analytic is that its truth can be known simply by analysing the meanings of all its constituent terms. What makes it a logically necessary truth is that to deny it would involve self-contradiction: its contradictory would be a logically necessary falsehood (2.22). Instead, and very reasonably, Horatio wanted a synthetic and logically contingent proposition; one the truth or falsity of which could not be known merely by fully understanding its meaning, and the contradictory of which would be neither necessarily true nor necessarily false. Had Hamlet claimed—however anachronistically, and irrelevantly to Shakespeare's dramatic purposes—that all Danish villains were the products of maternal deprivation; then his proposition would have been not analytic but synthetic, not necessary but contingent. It would also have been aposteriori, in as much as it could be known to be true—if it were true—only by reference to some actual study of the deprived home background of Danish villains. (I write both ' aposteriori ' and its opposite ' apriori ' as single unitalicized words deliberately. It is chauvinistically, you might even say bigotedly, purist to treat as unnaturalized aliens immigrants first landed at least a quarter of a millennium ago.)

3.6 In terms of the fundamental distinction explained in the previous two paragraphs (3.4 and 3.5), the No-true-Scotsman Move consists in responding to the falsification of a contingent proposition by covertly so reinterpreting the words in which it was originally formulated that these now become the expression of an arbitrarily made-to-measure necessary truth. This manoeuvre always involves either a high or a low redefinition of a crucial term: where the qualifications for membership of the class are increased, we have a high redefinition; and where they are reduced, we have a low redefinition. To rate as a true Scot you have to be not merely a Scot but a Scot who is not a sex maniac. To score—if that is the word—as one of the hated honkies you have to be not merely a Caucasian but a Caucasian who is not a Latin American Communist. What must have been an example of the low redefinition of the word ' socialist ' was provided by Queen Victoria's eldest son, the future King Edward VII, when he—of all people!—said in a speech at the Mansion House: ' We are all socialists nowadays ' (5/xi/1895).

3.7 The temptation of course is not just to slide, under the pressure of falsification, from a contingent to a logically necessary interpretation. It is to fail to recognize what has happened, and so to be apt to slide back again into the original interpretation immediately that pressure is removed. Our Scottish chauvinist is all too likely to go away believing that his original over-confidence in Scottish superiority has been vindicated; while Mr Carmichael's starkly racist picture of the world was probably preserved without the revision which he himself had, in his own way, admitted that the existence of the Castro regime made necessary. It is important to be aware of the possibility of analogous equivocations (1.5) even where we have not been alerted by a flagrant No-true-Scotsman Move.

3.8 Suppose, for instance, that someone assures us that all criminal behaviour, or all sexual deviation, or all suicidal desires and decisions, are symptomatic of mental disease. Such things frequently are said nowadays, and not least by people paid to know. The first question to press upon such psychiatric spokesmen, before we begin to ask whether what is said is true, is: ' What sort of thing actually is being said ? ' Are we really confronted by a genuine contingent claim, or is it all ultimately a matter of definition and of more or less arbitrarily chosen criteria ? Are they claiming, that is to say, to have made a discovery; the discovery perhaps that all those who commit crimes are also, by quite independent criteria, mentally diseased ? Or is it that they are taking the committing of a crime as itself a criterion of mental disease; insisting, in effect, that such behaviour constitutes a logically sufficient condition of being mentally diseased; and hence that for them it would be contradictory to say that a man is committing a crime and yet is not mentally diseased (2.16) ? Is it, in short, as it used to be when suicide was still under English law a crime; and juries, without hearing any expert psychiatric evidence, regularly returned the verdict, ' Suicide, while the balance of the mind was disturbed '? (See Flew (4), passim.)

3.9 The fact that usually the crucial distinction will not have been formulated by the other person makes it much more, rather than less, necessary to press the question raised in the previous paragraph. For until he has made this distinction, and made it sharply, it is bound to appear to him that the necessary truth interpretation of the words he is uttering provides unassailable support for the same words construed as reporting a psychiatric discovery. As a text-book illustration of the seductive confusion into which it is so easy to fall, consider an argument, current since 1651, when Thomas Hobbes published his *Leviathan*, if not from earlier still. The premise is that all actions must as such be motivated. But in that case, it is urged, they all are done, and could only be done, because the agent wants to do them. Since to do always exactly what you want to do is to be utterly selfish, it appears to follow that there is, and could be, no such thing as an unselfish action.

3.10 Wait a minute! The premise is presumably construed as analytic and necessary. Certainly no other grounds are offered why we should accept it as true. But the conclusion is, equally certainly, put forward as a scandalous and demoralizing revelation. This by itself is enough to show that there must be something wrong. It cannot be right to infer a substantial conclusion from a merely tautological premise. Exactly what is going wrong, and how, can best be seen by displaying the ambiguity of the crucial term ' want '. In the premise to want is simply to have a motive, any motive at all. It is only this equation which permits us to go on to say that all actions ' are done, and could only be done, because the agent wants to do them '. But in the conclusion ' to do always and only exactly what you want to do ' has to be taken in a different and stronger sense: the sense in which someone might claim to have gone to visit a sick relative although they did not at all want to do so. It is only in this interpretation that the proposed conclusion becomes exciting and surprising; and carries the shocking implication 'that there is, and could be, no such thing as an unselfish action'. So construed it is, however, not supported by the premise; and in any case, as my illustration must have suggested, it is known to be false.

3.11 Another rather more subtle example of such an alternation between radically different interpretations of the same or very similar forms of words is provided by Mrs Joan Rockwell's *Fact in Fiction*, boldly subtitled 'The use of literature in the systematic study of society'. In her Preface she writes: 'My basic premise is that literature neither "reflects" nor "arises from" society, but rather is an integral part of it, and should be recognized as being as much so as any institution, the Family for instance, or the State' (p. vii). So long as this is interpreted as, in effect, a matter of definition it is clearly only too true. But in her very next sentence the author slips towards a more substantial interpretation of this 'basic premise'. She asserts: 'Narrative fiction is an indicator, by its form and content, of the morphology [shape?] and nature of a society. . . .' In the main body of the book we find that first denial forgotten. It is no longer false but, it seems, necessarily true that literature 'reflects' the society within which it arises: 'To say that writers necessarily reflect their own times, which I must repeat is the justification for using their fictions to study the facts of their society, is to say that they are bound to do so, and cannot choose to do otherwise' (p. 119).

3.12 It is one thing, and quite indisputable, to say that anything which happens in society is, by definition, a social phenomenon: 'my basic premise is that literature . . . is an integral part of' the society in which it is produced. But it is quite another thing, and not one to be accepted without the deployment of a deal of evidence, to say that literary products always contain information about the societies in which they were produced: ' writers necessarily reflect their own times'. Maybe they do. Maybe it is impossible for even the best of historical novelists or of science fiction writers to cover their parochial social tracks. But to show this it is certainly not sufficient to appeal to the quasi-definitional truth that the production of literature is a social phenomenon.

3.13 The last nine paragraphs (3.4-3.12) have tried to explain the fundamental distinction between necessary and contingent propositions, and to indicate some of its importance. The No-true-Scotsman Move (3.1-3.3) is an attempt to evade falsification: a piece of sleight of mind replaces a logically contingent by a logically necessary proposition. We need now both to link these ideas with the Popperian insights described in the previous chapter (2.9-2.11) and to bring out why any such manoeuvre is inconsistent with a forthright concern for truth.

3.14 The heart of the matter is that the substance and extent of any assertion always is, and must be, exactly proportionate to the substance and extent of the corresponding denial necessarily involved in the making of that assertion. To assert a proposition is as such to deny the contradictory of that proposition (2.22). So the more you want to say, the more you have to stick out your neck. The wider and the more substantial your claims, and the greater the scope and the explanatory power of the theory which you propose; the greater must be the range of theoretically possible occurrences which would, if they actually occurred, falsify your claims or your theory. In the words of a fine Spanish proverb: ' " Take what you like ", said God, " take it; and pay for it." '

3.15 The No-true-Scotsman Move can thus be seen as one of too many alternative manoeuvres for trying to get hold of a big dollop of substantial truth without paying for it. The original contingent claim was wide and substantial. What it denied made it falsifiable by the misbehaviour of any Scot; and it actually was, we are supposing, falsified by the deplorable conduct of our imaginary Mr Angus MacSporran. The substitute assertion was not in the same way substantial. Precisely because it could not be falsified by any describable occurrences, it was not really making any assertion at all about what is or is not supposed to happen in the universe around us. This is what was meant when, in that enigmatic masterpiece the *Tractatus Logico-Philosophicus*, Ludwig Wittgenstein wrote: ' The propositions of logic are tautologies. The propositions of logic therefore say nothing. (They are the analytical propositions.) ' (§§ 6.1-6.11.)

3.16 The logical observation of the last paragraph but one (3.14) constitutes another of the foundations of the Popperian philosophy of science: ' The best we ever have, or ever could have, can, and could only be, the best so far ' (2.11). This rejection of all ultimacy in science, a rejection inspired especially by Einstein's overthrowing of the Newtonian establishment, is no licence to abandon the critical pursuit of truth: quite the reverse. Certainly science requires openness, and bold conjectures. But the openness required is openness to the possibility of new, and possibly upsetting discoveries of what actually is the case; while the conjectures must be conjectures about what is in fact true. And it is, as I have been urging, the mark and test of our sincere concern for truth that we should be in this way open to and ready to accept the falsification of our claims about how in the universe around us things are. By withdrawing your attention from flesh and blood Scots, to talk of true Scotsmen, you show that your concern now is with what you would like, rather than with how in truth things are.

3.17 It is these considerations which have led Popper to maintain that falsifiability is the criterion of demarcation between science and non-science. It is, that is to say, an essential mark of a scientific hypothesis that it should be in principle falsifiable; that there should be describable phenomena which, if they were to happen, would, by their actual occurrence, show that that hypothesis was false. A theoretical structure, like that of Newton, which is eventually shown to have been false, is not thereby shown to have been unscientific: ' We cannot identify science with truth, for we think that both Einstein's and Newton's theories belong to science, but they cannot both be true, and they may well both be false ' (Popper, quoted Magee, p. 28).

3.18 What must disqualify a theory, or a theoretician, as unscientific is, rather, that it, or he, refuses to allow for any things which if they were to occur, would constitute falsification. If a statement is to be substantial, then it has to deny something; something the past, present, or future occurrence of which would constitute, or would have constituted, falsification. And if a theory is to explain why this happens, then it must explain why it is this and not something else which happens. So when we learn from one of the classics of anthropology of all the many cunning intellectual devices by which the Zande ensure that no describable occurrence ever could constitute a falsification of their witchcraft beliefs, then this very discovery constitutes a sufficient reason for not awarding to this thought system the diploma title ' Zande science ' (Evans-Pritchard, passim.) And, as Popper himself argues, the fact that both Freud and Freudians seem always able and eager to show, after the event, that any and every apparently falsifying fact was after all what their own theories should have led us to expect, constitutes a very good reason for doubting whether, whatever else it may be, psychoanalysis is a science.

3.19 Either or both of these particular examples may be contested. But the general moral remains. Whenever we are uncertain, as certainly we often should be, how much if anything is actually being said, or whether we really are being offered a genuinely explanatory and scientific theory, then we ought to press home The Falsification Challenge: ' Just what, please, would have to happen, or to have happened, to show that this statement is false, or that this theory is mistaken?' (Flew and MacIntyre, pp. 96-100).

3.20 For the wife of that doomed soldier outside Stalingrad belief in God was—at least as far as the present test goes—scientific: ' If there is a God . . . then He will bring you back to me soon and healthy' (2.1). Her hypothesis was falsifiable; but in fact false. It is not equally obvious that the same could be said when that original religious hypothesis is enriched, but at the same time qualified, in the fashion of the father of Sir Edmund Gosse: ' Whatever you need, tell Him and He will grant it; if it is His will ' (Gosse, Chap. II). Fear of falsification has led many a brash, decently falsifiable hypothesis to be progressively eroded away—the death by a thousand qualifications.

4 Motives and Grounds

4.1 The No-true-Scotsman Move is one way in which ground may be shifted, often without anyone realizing what is going on. There are many others. Look at The But-they-will-never-agree Diversion. This is another type of instance in which bargaining considerations are intruded into factual discussion, where such considerations are totally irrelevant. If you are trying to thrash out some generally acceptable working compromise on how things are to be run, then you have to take account of the various sticking points of all concerned. But if instead you are inquiring into what is in fact the case and why, then that someone refuses to accept that this is true is neither here nor there. He may be right or in error, reasonable or unreasonable, in his refusal. But the questions which we were supposed to be discussing are not questions about him and what he does, or will, or will not, accept. They are a matter of what, regardless of what either he or anyone else may either wish or think, is true.

4.2 This But-they-will-never-agree Diversion is often executed with the help of the wretched expression ' prove to '. The point of this expression, and what makes it deplorable, is to confound producing a proof with persuading a person. Yet a man may be persuaded by an abominable argument; just as he may remain unconvinced by considerations which he certainly would accept if only he were more rational, or more honest, or both. It may very well be that no one can prove to—that is to say persuade—Soviet spokesmen that the tanks were sent into Hungary in 1956, into Czechoslovakia in 1968, and wherever next, in order to preserve the world's largest and most unyielding empire. Presumably most of the enthusiastic amateurs, if not the professionals, would cease to be spokesmen if they were to be so persuaded. But all this is simply irrelevant to the questions whether the original proposition is true, and known to be true.

4.3 The But-they-will-never-agree Diversion is one sort of move from whatever was the original subject of discussion to a quite different sort of question, a question about persons. Another—which I christen The Subject/Motive Shift—is far more common. This moves: from discussing the truth or falsity of some proposition, and the grounds for holding that it has this one or that one of these alternative truth values; to discussing the quite different questions of what someone's motives might be for asserting or denying the proposition, and or for wishing to believe or to reject it. Once this distinction is clearly made, it becomes obvious that these are indeed different kinds of question. But we still need, both to find ways of keeping them distinct, and to pick out some of the possible connections between the two.

4.4 One reason why they are so often confounded is that the word ' reason ' is itself relevantly ambiguous. When someone is said to have some reason for believing a certain proposition, we may need to ask: whether this reason is a ground for holding that that proposition is actually true; or whether it is a motive for persuading himself of it, irrespective of whether it is true or not. In the former case we can speak of a reason (ground), in the latter of a reason (motive). We might also distinguish a third sense. For in the phrase ' One reason why ', at the beginning of the present paragraph, what is being referred to is neither of these two but a reason (cause).

4.5 The classic occasion for distinguishing the first two of these three senses is provided by the famous argument known as Pascal's Wager. Others have urged that certain features of the universe constitute evidence for the existence of some kind of God: there is, for instance, the Argument to Design. But in the *Pensées* Pascal begins by conceding that there are no grounds adequate to warrant the desired conclusion. We are, he maintains, in what is essentially a gambling situation. It is one in which a prudent calculation of the inescapable stakes and the possible winnings must lead any sane man to try to persuade himself of the truth of the Catholic faith, notwithstanding the alleged complete inadequacy of the evidence. God or no God? No man alive has any choice but to stake his life. Upon each man's throw there hangs for him heaven, or hell, or annihilation. Pascal urges: ' If you win, you win the lot. If you lose, you lose nothing. So . . . bet on the existence of God.' Clearly, in terms of the present distinction, Pascal is saying that, though we have no good reasons (grounds) for believing, we have the very best of reasons (motives) for trying to persuade ourselves. (See Flew (3), VI 7; and compare Flew (2), passim.)

4.6 The further distinction between reasons (grounds) and reasons (causes) becomes essential if we want to tackle an argument such as the following: ' In deriving mind and knowledge from nature, as science conceives it, " the naturalist " must assume that his own account of nature is true. But on his premises, the truth of this account, like that of any other bit of knowledge, is merely the function of the adjustment of the organism to its environment. . . . This entire conception of knowledge refutes itself ' (Urban, p. 236). This writer, like so many others, is, in effect, arguing: that if there are always physiological reasons why I utter the sounds which I do utter; then I cannot have, and know that I have, good reasons for believing the propositions which I assert by making those sounds. If it is the one, then it must be merely that and not the other. But now, in terms of our further distinction, the physiological reasons must be reasons (causes), whereas the good reasons which I may or may not have for believing can only be reasons (grounds).

So no reason (ground) has been given why reasons, in these two different senses of ' reason ' should be taken as being necessarily rivals for the same space, with the presence of one precluding that of the other.

4.7 The first general lesson of method for us to draw from the last three paragraphs (4.4-4.6) is that, wherever we need to distinguish two or more senses of a word, there we also need to supply informative parentheses in order to maintain the distinctions made. The model to imitate, which I have been imitating, is that of funny (ha ha) as popularly and correctly distinguished from funny (peculiar).

4.8 A second general lesson perhaps begins to come out only when we work—as here—with engagingly important and tricky examples. It is that it can be hard fully to come to terms with the fact that a word may be entirely ambiguous; with senses as different, and as unconnected logically, as those of any two wholly different and etymologically unrelated words.

4.9 The temptation is to plunge on as if the crucial distinctions had never·been made. Thus, while claiming to have accepted the distinctions indicated earlier (4.4), someone may nevertheless continue to argue just as before: if the reason why I believe is physiological; then I cannot have any sufficient reason for believing. Or take quite another example. Suppose that it has been conceded that there are two radically different senses of the word ' democracy '. In one of these two senses—call it liberal—an institution is democratic to the extent that it is in due season possible for its members to vote the leadership out, if that is what those members themselves want to do. In the other sense—call it paternalist—an institution is democratic to the extent that it serves the true needs and interests of its rank and file members, perhaps as determined by somebody else. It is tempting to speak now of two varieties or species of democracy—especially if you happen to want to appropriate for one of these, favourable attitudes originally directed towards the other. Yet, precisely to the extent that the two senses of the word ' democracy ' are different, this must be wrong. For, just to that extent, there is no common genus of which the two supposed species or varieties can be species or varieties.

4.10 The reason why these mistakes of perseveration are so tempting—the cause, that is—is that speech habits are just as much habits as any others; and just as hard to make or to break. Maintaining a distinction between two senses of a word requires that some entrenched habits of association be overcome, and that others be formed. The least which one can do is to make suitable parenthetical insertions whenever we employ the ambiguous word in any possibly troublesome context. In this way the desired associations of what is inserted pull against the undesired associations of the ambiguous term. In the exceptionally intractable case, where this routine treatment proves ineffective, the ambiguous term should be jettisoned completely, and replaced by two words or expressions which do not look alike, and so do not have the same associations.

4.11 The most common, indeed the most dully common-place, case of The Subject/Motive Shift is that in which an assertion is dismissed as false, or an argument is discredited as unsound, simply because it is made, or presented by, an interested party. It may very well be that as an Ambassador, or as a Trades Union Official, or as a Public Relations Officer of some business firm, he is paid to say that, or to argue in this way. But that interest by itself is not a sufficient, or indeed any, ground for concluding that his claims must be false and his arguments invalid. You and I too, who so far as I know do not belong to any of these three suspect classes, often present sound arguments for conclusions which happen to be both true and to our advantage. The truth is no more necessarily disagreeable than it is necessarily agreeable.

4.12 No one, of course, would outright deny the banal remarks of the previous paragraph. The problem is, as

with many other important truisms, so to discipline ourselves that we are never carried away by the heat of the
moment. Even a truism, we have always to remember, may
be true. One step in the right direction might be for everyone to collect their own sets of arresting examples. Such
sets need to be individually collected and individually
weighted so as to offset each individual's particular assem-
blage of biasses. If you yourself are generally inclined to
suspect, for instance, businessmen, or diplomats, or labour
spokesmen; then your personal collection has to contain
statements from members of these particular, notoriously
sinister, occupational groups. And these must be statements
which were, when made contemptuously, disbelieved by you
and by other opponents, but which nevertheless turned out
to have contained the truth and nothing but the truth.

4.13 The first of the last two paragraphs brought out that
The Subject/Motive Shift is fallacious, while the second
suggested a salutary disciplinary drill (4.11 and 4.12). But
this coarsely fallacious move, The Subject/Motive Shift,
needs to be distinguished from two others which are perfectly legitimate. Certainly it is fallacious to urge: that,
simply because someone has some sort of vested interest in
the truth of a proposition, or in the validity of an argument;
therefore, that proposition, or that argument, is, or very
probably is, false or, as the case may be, invalid.

4.14 Now suppose that that proposition is being assessed
as a piece of testimony. It remains relevant that a proposition may be both true, and known by the person who
asserts it, notwithstanding that person has every kind of
powerful vested interest in both its truth and the assertion
of its truth. But that it is asserted by someone who is in a
position to know, and has no reason (motive) for trying to
deceive us, is for us, who are not in his position to know,
better evidence for believing that it is true than the same
assertion made by someone in an equally good position to
know, but with opposite interests. There are, of course,
other dimensions of complexity in the reasonable assessment

of testimony. The present point is, however, simple and uncontroversial. It is taken for granted, for instance, by everyone who would in a question of foul play be inclined, all other things being equal, to accept the testimony of a neutral spectator rather than that of any of the contestants.

4.15 Another thing which is in itself perfectly legitimate also needs to be distinguished from the fallacious Subject/ Motive Shift. It is perfectly legitimate, at least from the standpoint of sound thinking, to raise and to pursue questions about interests and motivations. In particular it is innocuous, and it can be illuminating, to do this when the original issues of truth and validity have been settled. For, given that what he says is both mistaken and now known to be mistaken, and given that he is no fool, then we may reasonably ask what it is which is misleading him. Yet it will not do—notwithstanding that it is all too often done— to offer more or less speculative answers to such consequential questions as a substitute for, rather than as a supplement to, the direct examination of whatever were the prior issues.

4.16 It appears that this offence is more prevalent nowadays than ever before. Certainly the temptation has been much increased by the proliferation of psychoanalysis and of the sociology of knowledge. Consider a statement by a leading British Freudian: ' To achieve success the analyst must above all be an analyst. That is to say he must know positively that all human emotional reactions, all human judgements, and even reason itself, are nothing but the tools of the unconscious; that such seemingly acute convictions which an intelligent person like this possesses are but the inevitable effect of causes which he buried within the unconscious levels of his psyche ' (Berg, p. 190).

4.17 Suppose that the scope of this statement had been explicitly limited to the analytic situation, and that its application had been emphatically confined to neurotic patients crying out for psychiatric attention. Still, many lay readers would have taken it that psychoanalysis licenses the absolutely general conclusion 'that all human emotional reactions, all human judgements, and even reason itself, are nothing but the tools of the unconscious'. As it is it seems that Dr Berg himself is presenting this reckless claim as either a finding or a presupposition of psychoanalytic investigations. What else could be the point of his saying that this is something which the successful analyst 'must know positively'? Presumably some rule of professional procedure—never to attend to the truth or validity value of anything which the patient says in the analytic hour, but always to seek some motive for his saying it—is being confused with, and mistaken to warrant, the conclusion that such utterances—and hence, by a somewhat drastic extension, all utterances—do not actually have, and or cannot be known actually to have, any truth or validity value at all. Nothing like leather becomes, not for the first nor for the only time, nothing but leather.

4.18 Be it as it may with these interpretative conjectures. Certainly there is no doubt that, if psychoanalysis really does carry any such universal consequence, then the entire enterprise must thereby discredit itself. For this is exactly that same disastrous consequence which, as we saw earlier (4.6), has often, but wrongly, been said to be implicit in any form of scientific 'naturalism'. Let us put this objection in the most pointed and personal way. If 'all human judgements ... are nothing but the tools of the unconscious'; and if the point of this 'nothing but' is, as it surely must be, to preclude the possibility of their also being known to be true: then this must apply equally to the no more than human analyst, and that 'all' has to include the judgement in which it itself occurs. What is poison for the goose is poison for the gander; and for the farmer too.

4.19 It is, therefore, understandable, if not on that account venial, that usually those who see the putative insights of psychoanalysis as instruments for the discrediting of opinions, apply these instruments only to other people.

To lose the initiative would be fatal. It has to be the opinions of other people rather than our own which are, with no good grounds given, dismissed as nothing but the expressions of unconscious motivation. It has to be the motives of other people rather than our own to which The Subject/Motive Shift shifts. With appropriate alterations the same applies to the sociology of knowledge. It has to be the opinions of other people which are as such repudiated as nothing but the ' false consciousness ' of a particular social or economic class. Ours, naturally, are not in the same way mere ideology, but instead science; perhaps even ' scientific socialism '.

4.20 For anyone who sincerely wants to know what's what, the right moral points in the opposite direction. It is, that whatever insights psychoanalysis and sociology may be able to give us should be applied in the first resort to ourselves. It is most salutary to remember, and to follow, an example set by Charles Darwin. Long before Freud was even born Darwin made it his practice to note down all objections to his theories the moment he met them. That observant naturalist had noticed that we are all more apt to forget what we have some interest in not remembering.

4.21 Another common move which resembles The Subject/Motive Shift is not the same, although it too is unsound. Someone ripostes to a speaker, who has put forward a view which is distinctively Christian, or Marxist, or Conservative, or whatever: ' You would believe that, because you are a Christian, or a Marxist, or a Conservative, or a whatever '. Such remarks may often have point and value. But they cannot be admitted as objections. For that some view happens to be an essential element in some general position only begins to be relevant to the question of the truth of that particular view, when the general position is already known to be in error. But to appeal to this assumption of error in a debate with someone who starts as an adherent of that particular general position, constitutes a textbook example of Begging the Question: that is, of taking for granted precisely what is in dispute. And, furthermore, even if the general Christian, or Marxist, or Conservative, or whatever, position is in error it could—and almost certainly does—contain some elements and carry some consequences which are in their own right true.

4.22 Such attempts to refute a view merely by classifying it in some irrelevant way are not the same as attempts to refute it by referring to the possible motives of its protagonists. One favourite contemporary token of the former type is to meet a contention by dismissing it on no other or better ground that that it is well worn, or hackneyed, or boring, or predictable (1.46). It ought to go without saying that none of these facts has by itself any bearing at all upon the question of the truth value of the proposition so characterized. It is a scandal of our trendocratic times that a purely journalistic classification of this kind should so often be allowed to pass as a refutation.

4.23 Another token of the same type was common in the USA during the Vietnam War. Some said that if South Vietnam was allowed to fall to the Communists, then Laos, Cambodia, and other countries of South East Asia, would follow in quick succession. Their position was regularly described as The Domino Theory. Yet the very aptness of this happy description was almost equally regularly mistaken for a disproof. Certainly almost everyone whom I ever asked for reasons for thus dismissing this contention appeared to be startled by my request. Let us do our bit to discourage bad practice, by introducing an opprobrious label: The Fallacy of Pseudo-refuting Description.

5 Minding Our Language

5.1 The Fallacy of Pseudo-refuting Description is essentially obscurantist. For the effect, and too often the object, of committing this fallacy is to dismiss whatever is so described blindly and for no given reasons. For instance: the facts, or the alleged facts, that various sorts of Marxist-Leninists and their fellow-travellers have recently been securing more and more of the offices in Britain's powerful trades unions, and that they are in consequence exercising more and more influence over the formation of the policies of the British Labour Party, may, or may not, be welcome. But, quite certainly, it is obscurantist to prevent critical examination of these claims, and of their possible implications, by chorussing the catch-phrase ' reds under the bed '.

5.2 The employment in this way of such mindless but memorable catch-phrases constitutes a partial and piecemeal approach to the Newspeak of George Orwell's last appalling nightmare *1984*. Newspeak was, it will be remembered, the artificial language being developed in hopes that it would finally replace ' Oldspeak (or Standard English, as we should call it) by about the year 2050 '. The aim, as Orwell goes on to explain in his Appendix on ' The Principles of Newspeak ', was ' to meet the ideological needs of Ingsoc, or English Socialism. . . . Newspeak was not only to provide a medium of expression for the world-view and mental habits proper to the devotees of Ingsoc, but to make all other modes of thought impossible. . . . Newspeak was designed not to extend but to diminish the range of thought, and this purpose was indirectly assisted by cutting the choice of words down to a minimum.' He quotes a well-known passage from the American Declaration of Independence, beginning: ' We hold these truths to be self-evident, that all men are created equal, that they are endowed by their Creator with certain unalienable rights. . .'. The whole passage could in Newspeak be rendered only as ' doubleplus ungood crimethink '—an impoverishing misrepresentation, which carries with it a simultaneous and wholly prejudicial shudder of rejection (1.56).

5.3 Orwell's concern with language as the main vehicle of thought, and his commitment to struggle against all the tendencies which the inventors of Newspeak laboured to promote, can also be seen in his splendid essay on ' Politics and the English Language '. This concern and this commitment is not the same as, although it does often accompany, concern for and commitment to euphony and literary elegance. Both are of course often found together—in Orwell himself, for instance, and again in the classic handbooks of Sir Ernest Gowers. But any preference which we may have for these Old World monosyllables ' car ' and ' lift ', as opposed to their less terse translantic equivalents ' automobile ' and ' elevator ', is, when we are thinking about thinking, neither here nor there. What by contrast has to be relevant is any usage or abusage which tends either to reduce the stock of concepts available or to conceal the meaning of what is supposed to be being said.

5.4 We might, I suppose, in a philistine moment dismiss issues of the former kind as merely, that is to say trivially, verbal. But, if these are to be our paradigms of the merely verbal, then it becomes preposterous to condemn upon the same grounds questions of the second kind too. For the nub of the distinction between these two different sorts of issue about words precisely is: that the one refers only to possible forms of expression, and not to the content of what might be said; while the other is concerned essentially with the meanings which are or might be expressed. The question whether we should say ' He got in touch with her ', rather than ' He contacted her ', is indeed trivially verbal. But the question whether we should say ' He did kill him ' rather than ' He did not kill him ' is, notwithstanding that it can be represented as a matter of whether or not to insert one particular three-letter word, a matter of substance. It happens to be, in the most literal sense, a question of life and death.

5.5. It was, therefore, though witty, unfair of Edward Gibbon to ridicule the Christian world for splitting over an iota. For the fact that the dispute between those contending that the Son is of like substance (Greek, *homoiousios*), and those maintaining that He is of the same substance (Greek, *homoousios*), can be represented as a dispute over the insertion or excision of one little letter, is a wretched reason for suggesting that any difference so symbolized must be correspondingly insignificant. It is of course none of our present business to decide whether this particular conclusion happens to be true, or whether the great conflict centred on a distinction without a difference, or whether it all was in some other way misguided or gratuitous. For us here the point is simply that Gibbon's witticism provides no support whatever for any such conclusion.

5.6 Right from the beginning of Chapter One I have been pointing out that and why, if we want to think better and straighter, we have to school ourselves to follow stricter and perhaps less common usages of certain crucial terms: ' valid ' and ' invalid ', for instance, ' true ' and ' false ', ' know ', ' refute ', and ' prejudice '. Orwell, in the essay mentioned in the last paragraph but two (5.3), extends this sort of concern about language to include style and syntax as well as vocabulary: ' People are imprisoned for years without trial, or shot in the back of the neck, or sent to die of scurvy in Arctic lumber camps: this is called " elimination of unreliable elements ". . . . The inflated style is itself a kind of euphemism. A mass of Latin words falls upon the facts like soft snow, blurring the outlines and covering up all the details. The great enemy of clear language is insincerity. When there is a gap between one's real and one's declared aims, one turns as it were instinctively to long words and exhausted idioms, like a cuttlefish squirting out ink.'

5.7 Fuller development of the theme that some improvements of our style are necessary to the improvement of our thought I must leave to others: Orwell himself, for instance, and Sir Ernest Gowers. It is enough here for me to have said that this is so, and to have insisted that this kind of concern about language is not to be dismissed as trivially verbal, whereas another might be. Orwell also suggests that the call for clear, brief, careful, concrete and down-to-earth expression is a call for integrity, for honesty both to oneself and to other people. So let us write down into our copy books two of the maxims of that great French aphorist, the Marquis de Vauvenargues: ' Obscurity is the kingdom of error '; and, ' For the philosopher clarity is a matter of good faith.'

5.8 Vagueness too can be a fault. The first thing to notice about vagueness is that it ought not to be, although it often is, equated with ambiguity. To say of some word or expression, or of some whole statement, that it is ambiguous is to say that it can be construed in two, or at least two, quite different ways. (See, for instance, 1.14-1.15 and 4.5-4.7.) Anyone who says this asks for, and should be ready to meet, the challenge: ' What two interpretations do you want to distinguish ? '

5.9 To complain of vagueness is to complain that what has been said is, in some relevant dimension, unacceptably indeterminate. To promise to come round during the afternoon is to undertake a less precise commitment than to undertake to arrive between sixteen and seventeen hundred hours. Neither the vaguer nor the more precise of these alternative undertakings is in any obvious way ambiguous. Equally you may say something ambiguous without either of its alternative interpretations being such as to expose you to the charge of imprecision. Suppose you ask me: ' How did you expect them to behave? ' Then your question will be importantly ambiguous. For it might reasonably be construed as an inquiry: either about what I had believed that they would in fact do; or about what I had considered that they ought to have done. Yet in neither of these quite different interpretations would your question have been in any obvious way vague or imprecise. Since this ambiguity of the word ' expect ' is one which we need to keep in mind, I must follow my own excellent advice (4.7). Let the first of the two senses just distinguished be labelled descriptive and the second prescriptive.

5.10 The second thing to notice about vagueness is that it is not a fault in a language that it provides, or provides for, some words and expressions which are vaguer than others. It is reasonable to complain only when and in so far as ' what has been said is, in some relevant dimension, unacceptably indeterminate '. Suppose that I am in no position to say more about my time of arrival than that it will be during the afternoon. Then it must be unjust to condemn me for not specifying closer limits; and it would be an awkward defect, not a merit, in my language if it rendered me unable to say the only thing which I was entitled to assert. The traditional stock example of a vague term is the word ' bald '. No doubt we could so redefine this good old word that in its future correct usage it implied some specific density of, or some specific total, hair population. But this would be a silly move. For we should be exchanging a humbly serviceable tool, which we do often have occasion to employ, for a shiny new piece of futile equipment, which in our normal everyday life we never should be in a position to use correctly.

5.11 Even ambiguity may be innocuous. Little harm has ever come, surely, from the ambiguity of the word ' bank '? In one sense a bank is whereon, if both passers-by and various public authorities permit, Shakespeare's wild thyme blows. In the other a bank is what keeps your money, and understands, or claims to understand, students, or businessmen, or housewives, or whoever next. Sometimes too ambiguity may be exploited legitimately. There was, for instance, nothing improper before Trafalgar about that most famous employment of the word ' expect ', simultaneously yet consistently in both its prescriptive and its descriptive senses: ' England expects every man to do his duty.'

5.12 Ambiguity becomes dangerous only where both interpretations are relevant to the same context; and where there is, therefore, a live possibility of confusion. Once upon a time there used to be warm debates about the place of Religious Knowledge in the curriculum of British schools. The expression ' Religious Knowledge ' was used in both weak and strong senses. In the former it referred to knowledge of what religious beliefs are held; and what religious practices are pursued. In the latter it carried also the implication that these beliefs themselves constitute items of knowledge; and that these practices are, consequently, imperative. The danger, often realized, was that those who had failed to observe this crucial distinction would mistake it, that considerations sufficient to justify the teaching of Religious Knowledge (weak) must also, by the same token, be sufficient to justify the teaching of Religious Knowledge (strong). This cannot be right. For to justify the former it is not, while to justify the latter it is, necessary to proffer adequate reasons—whether these be grounds or only motives (4.4-4.6)—for holding that the religious doctrines are actually true.

5.13 Of course vagueness, in so far as it is a matter of ' what has been said ' being, ' in some relevant dimension, unacceptably indeterminate ', must be a bad thing. But, whether or not some particular measure of indeterminacy is in some particular context culpable, we have to notice a third general point. It is that any indeterminacy in the premises of an argument is bound to infect whatever conclusions can be validly deduced from these premises with a precisely corresponding indeterminacy. Given appropriate alterations exactly the same applies when it is a question of unreliability in the premises. That these things must be so follows from the essential nature of a valid deductive argument (1.10). Since to deny the conclusions of such an argument while nevertheless affirming its premises would be to contradict yourself, these conclusions must be implicitly or explicitly contained in those premises. It cannot but be, therefore, impossible to get out more or better than you put in. The situation is very like, indeed the same as, that epitomized in a slogan once popular among hard-bitten American computer people: ' Garbage produces garbage! '.

5.14 The fourth point to be made about vagueness is again one which also applies more widely. It is that if you choose to redefine some vague term in such a way that it becomes in the usage thus stipulated more precise, then you will have given another and different meaning to that word:

To remove vagueness is to outline the penumbra of a shadow. The line is there after we have drawn it, but not before.'

5.15 The remarks of the previous two paragraphs ought by now to appear obvious. But obviousness is not something always to be despised. There are moments when we need to remember the First Maxim for Balliol Men: ' Even a truism may be true.' Obviousness too really is, as truth and validity most emphatically are not, a matter of how it happens to seem to a particular person or group of persons at some particular time or times. What seems obvious to you may well not seem obvious to me; and what once to him appeared obviously untrue may now appear, even to him, an equally obvious truth.

5.16 Certainly there is no doubt but that able people do sometimes try to deduce conclusions both more precise and more reliable than their original data. Or take the philosophical problem of mind and body—the problem, that is, of the logical relations between talk about consciousness and

talk about stuff. Certainly able people have attempted t
resolve this by proposing so to redefine the word ' exper:
ence ', which at present carries an essential reference t
consciousness, that it would refer only to bodily movement:
with no embarrassing reference to consciousness at all. Ye
this proposal only begins to look sensible if we mistak
it that one and the same word ' experience ' could carry
simultaneously, both its present and a new and incom
patible sense; and hence that experience, which essentiall
involves consciousness, could be reduced to mere bodil
movements, which do not.

5.17 It is time to say something more about what ca:
and cannot be achieved by definition. First, it is just worth
saying that there is no call to try to define every term
Definition of one word in terms of other words can b
profitable only in so far as there are other words sufficientl
understood already. To demand either a definition or an
other kind of explanation where there is no relevant con
fusion or uncertainty to be removed is tiresome and
obstructive. There could be no better authority here thar
Dr. Johnson: ' Sometimes things may be made darker by
definition. I see a cow; I define . . . " Animal quadrupe:
ruminans cornutum ". But a goat ruminates, and a cov
may have no horns. " Cow " is plainer ' (7/iv/78). I thinl
too of a *New Yorker* cartoon which showed indignant wife
confronting contentedly drunken husband: ' No, I will no
begin by defining " soused "! '

5.18 Nor, second, is the ability to produce an adequate
definition a necessary condition for possessing a sufficien:
understanding of the meaning of a term. It is tempting, bu:
wrong, to argue, with the Socrates of Plato's *Republic*, that
' if I do not know what justice is, I can scarcely know
whether it is a virtue or not, and whether its possessor is or i:
not happy ' (§354C). For you may be able to employ some
word correctly and with understanding on all ordinary
occasions without being able to respond to such a ' What i:
—? ' question with a formal definition. It is a point made

n his own way by St. Augustine in his *Confessions*: ' And I confess to thee, O Lord, that I still do not know what time is, while . . . I know that I am saying this in time ' (XI (xxv)). Certainly this venial ignorance did not stop him, and will not stop us, from employing ordinary temporal expressions correctly and with understanding. We all know perfectly well what he meant when, in his unregenerate days, he prayed: ' O give me chastity; but not just yet! '.

5.19 The moral is that we need definition or other explication only where there already is, or where we may reasonably expect that there will be, some relevant confusion or uncertainty about meaning. A third point is that in such suitable cases for treatment the right prescription is not always a formal definition: beginning ' X is . . . ' or, better, ' The word " x " means . . .' Suppose that someone is going on about alienation. If he gives us a definition this will most likely be in terms of other theoretical notions derived from the young Karl Marx. But what is really required is some indication of how this theory engages with, and of how its truth value might be determined by reference to, what actually happens. Hence the illuminating question is: not, ' How would you define the word " alienation "? '; but, ' How could you tell that a person was not, or was no longer, alienated? ' When, but only when, we are equipped with workable criteria for alienation can we entertain such theories as a possible contribution to sociology as a science (3.15-3.18).

5.20 A fourth point, already suggested in the previous three paragraphs, is that definitions are of words or of their meanings, and not of whatever the words might be used to refer to. It is good practice always to bring this out by putting both the word or expression to be defined (dog Latin, *the definiendum*) and the supposed or proposed equivalent (correspondingly, *the definiens*) between inverted commas. The act of definition thus consists in saying either that the definiendum is, or that it is to be, equivalent in meaning to the definiens. Where the claim is that by

established correct usage the two already are equivalen'
we speak of a descriptive definition. Where some innovatio
is involved, we speak of a prescriptive or of a stipulativ
definition.

5.21 Some people who grasp our fourth point too wel
may, like Humpty Dumpty, be carried away by the possi
bilities of stipulative redefinition. Remember that ofte
quoted exchange from Chapter Six of Lewis Carroll'
Through the Looking Glass: ' " I don't know what you mea
by ' glory '," Alice said. Humpty Dumpty smiled con
temptuously. " Of course you don't—till I tell you.
meant ' there's a nice knock-down argument for you! ' '
" But ' glory ' doesn't mean ' a nice knock-down argu
ment '," Alice objected. " When I use a word," Humpt
Dumpty said in rather a scornful tone, " it means just wha
I choose it to mean—neither more nor less." " The questio
is," said Alice, " whether you can make words mean s
many different things." " The question is," said Humpt
Dumpty, " which is to be master—that's all." '

5.22 Certainly it is a matter of human choice, rathe
than of natural law, that in particular languages particula
sounds and shapes have the meanings which they do have
Yet this fact of ultimate and collective human decision mus
not be mistaken to imply that it is either right or possibl
for any of us individually to endow any such shape or soun
with whatever meaning we may happen to wish on it. I
was all very well for whoever first introduced the concept o
gas to give that sense of his choice to the previously un
employed monosyllable ' gas '. It is a different thin
altogether to announce that you now propose to use som
already serving word in a sense quite other than that b
usage established. Again I quote Dr Johnson, although h
brings in the rather special case of names: ' My name migh
have been " Nicholson " originally as well as " Johnson "
but if you were to call me Nicholson now, you would cal
me very absurdly ' (20/vii/63).

5.23 To do what Humpty Dumpty did, and what to

many real people also do, is not merely to speak ' very absurdly '. It is also to act in bad faith. To express oneself in a public language is to undertake to speak, and to ask to be understood, in accordance with the established meaning conventions of that language. So to say something in some public language, and then afterwards to insist that you intended it to be interpreted in accordance with some private and previously unexplained conventions of your own, is to break the contract which you implicitly made when you started to speak in that language.

5.24 The fact that it is dishonest to make without due notice substantial departures from accepted usage, is no reason to rule out all stipulative redefinition. Like all other human institutions languages can be made better, as well as worse, by deliberate acts of policy. But reformers do need to take full account of the facts that linguistic habits are habits; and that, like other habits, they take a lot of changing (4.10). It is one thing to stipulate a meaning for a fresh term: this is to ask us to acquire the corresponding fresh verbal habits. It is quite another to prescribe a change in the meaning of a familiar word; this is to demand not only that we learn new skills and new associations but also that we unlearn old ones. This unlearning can be both hard and slow.

5.25 If, therefore, we want in a quite straightforward way to advance understanding and to communicate, we shall not propose new senses going right against the grain of current usage. Any redefinitions of familiar everyday words will call for modest redirections, not revolutionary reversals. One classic object lesson in what not to do is provided by the American sociologist Thorstein Veblen. In *The Theory of the Leisure Class* he introduces the expression ' conspicuous waste ' in a technical sense, stipulating most emphatically that in his book it is not to carry the customary overtones of condemnation. As he ought to have realized, and perhaps did, such old and strong associations cannot, and will not, be broken ' at a stroke '. It soon becomes

clear that he himself, like his readers, continues to constru[e]
'conspicuous waste' in an unfavourable sense (Veblen[,]
pp. 97 ff.). What pretends to be neutral sociologica[l]
science thus becomes in fact committed sociologica[l]
moralizing.

5.26 Mention of Veblen's pseudo-scientific and possibl[y]
disingenuous employment of the expression 'conspicuou[s]
waste' leads us to notice that many words and phrases carr[y]
such built-in value commitments. Since Aristotle first mad[e]
this observation—in his *Nicomachean Ethics* (1107A 9-12)—
literary critics, psychologists, linguists, philosophers, an[d]
others, have given a deal of attention to this fact, and hav[e]
developed various distinctions: between neutral or purel[y]
descriptive meaning, on the one hand; and, on the other[,]
expressive, emotive, normative, or evaluational. In a ver[y]
short and introductory book I cannot afford to say muc[h]
about this. But there are three points which absolutel[y]
must be made.

5.27 The first is that there is nothing wrong with th[e]
second sort of meaning, nor with the words whose meanin[g]
is partly or wholly of this sort, as such. If there were i[t]
would, paradoxically, be wrong to say so. For the wor[d]
'wrong' is obviously one which must itself fall under an[y]
such all-inclusive embargo. Of course, excessive employment of incendiary terms will prevent or disrupt rationa[l]
discussion of the issues so presented. Heat is frequently th[e]
enemy of light. But to prohibit all purely or partly evalua-
tive language is to put every form of valuation beyond th[e]
pale of rationality. The now popular misconception tha[t]
this is how things are, or ought to be, does a lot to brin[g]
about a state in which, in the often quoted words of Yeats[:]

> The best lack all conviction, while the worst
> Are full of passionate intensity.

5.28 The truth is that it is neither relevant nor sensibl[e] to object that discourse contains emotive words and value judgements, if its object just was to establish conclusions about what attitudes ought to be adopted or what action ought to be taken. What is offensive to reason, and what does constitute a ground for objection, is any arbitrariness in the application of these words, or any failure to deploy appropriate good reasons for whatever such judgements are in fact made. Veblen's proceedings were not obnoxious because he was denouncing conspicuous waste. Waste, whether conspicuous or not, ought to be denounced; and, even better, stopped. What was wrong was that Veblen was in fact denouncing conspicuous waste while pretending to be a neutral scientist. The point was: not that there should not be sociological moralists; but that no one has any business, when he is acting in that quite different capacity, to wear his scientist's cloak.

5.29 The second point about this second sort of meaning concerns arbitrariness. Everyone can conjugate the highly irregular verb which is supposed to run: ' I am firm ', ' You are obstinate ', ' He is pig-headed '. It is easy, and consequently common, to draw the diametrically wrong moral. The temptation is to construe this wayward con- jugation as supporting a general theory put forward by the always bold, not to say reckless, Thomas Hobbes. In 1651 he suggested in his *Leviathan* that all terms which carry any meaning of this second sort are to be defined with reference to the speaker: ' . . . whatsoever is the object of any man's appetite or desire, that is it which he for his part calleth " Good "; and the object of his hate and aversion " Evil ", and of his contempt " Vile " and " Inconsiderable ". For these words " Good ", " Evil " and " Contemptible " are ever used with relation to the person that useth them . . . ' (Chapter 6).

5.30 This particular essay in radical philosophy is bound to appeal, both to the cynic in all of us, and to the secular conviction that value is not a natural fact independent of the existence of the human race, but is somehow a projection of human inclinations and human aversions. But cynicism is not always, what it always claims to be, realism. And we have already seen, in the surely similar case of meaning, that to say that something is not a natural fact independent of all human desire and human choice is not immediately to license the conclusion that it is a creature of any and every individual caprice (5.21–5.22).

5.31 The crucial objection to this first Hobbist theory was put by David Hume, exactly a century later, in his *Inquiry concerning the Principles of Morals*. (Since Hobbes also offered another more social theory, according to which value must be the creature of sovereign power, we could call the present suggestion his Humpty Dumpty thesis.) No one could have been more convinced than Hume of the fundamental secular point that value is not a natural fact independent of all men. Yet he insisted on a distinction: 'When a man denominates another his " enemy ", his " rival ", his " antagonist ", his " adversary ", he is understood to speak the language of self-love, and to express sentiments, peculiar to himself, and arising from his particular circumstances and situation. But when he bestows on any man the epithets of " vicious " or " odious " or " depraved ", he then speaks another language, and expresses sentiments in which he expects all his audience are to concur with him ' (IX (i)).

5.32 Because Hobbes and Hume express themselves in the idiom and the prose rhythms of their different centuries both passages may require, as they will certainly reward, a second reading. The upshot is that it must be wrong to abuse words and phrases from Hume's second class by treating them as if they belonged to his first class—the class to which Hobbes apparently wanted to consign all valuing words. Suppose we allow, as we surely must, that all the phases of the supposed irregular verb ' I am firm ' belong in Hume's second category. Then the man who commends his own conduct as firm, when he would condemn the same conduct in a third person as pig-headed, is being arbitrary. His arbitrariness consists in discriminating between two relevant cases for no good reason. He is like the woman who indignantly denied the charge of hoarding, on the grounds that she herself was only taking care to stock up before the hoarders got everything.

5.33 The conclusion to draw is, therefore, not Hobbist but Humean. Sincere debate about values, and in particular about morals, largely consists in the discovery and removal of such arbitrary discriminations. It is, furthermore, an important part of what it means to maintain that some protest, or stand, or attitude, is moral—as opposed to merely personal or partisan—that it appeals to principles and that the principles to which it appeals are to be applied consistently and impartially, if not universally. Everyone knows why we impugn the sincerity of the selective moralist; or, better, ' moralist '. He professes a moral objection to— say—the use of poison gas. But when gas is being used by some party which he happens to favour—CS by white, capitalist, Americans against Communists in Vietnam or Lewisite by brownish, socialist Egyptians against Royalists in the Yemen—then, as the Greek tragedians used to say, a great ox sits on his tongue.

5.34 The third thing to notice in connection with the second sort of meaning is the phenomenon of Persuasive Definition. The phenomenon was first discussed extensively, and this label introduced, by the Michigan philosopher C. L. Stevenson. He distinguished between descriptive and emotive meaning, using ' emotive ' in an extremely comprehensive, ragbag sense. Persuasive definition consists in the attempt to annex either the favourable or the unfavourable emotive meaning of a word to some different descriptive meaning. Adolf Hitler was engaged in persuasive definition when he proclaimed: 'National Socialism is true democracy '. So, supposedly at the opposite end of the political spectrum, were those who christened the Soviet zone of

Germany 'The German Democratic Republic', and argued for their description (4.9).

5.35 Perhaps the best explanation, and illustration, is a passage which Stevenson himself quotes from Aldous Huxley's *Eyeless in Gaza*: ' " But if you want to be free, you've got to be a prisoner. It's the condition of freedom—true freedom." "True freedom!" Anthony repeated in the parody of a clerical voice. " I always love that kind of argument. The contrary of a thing isn't its contrary; oh dear me no! It's the thing itself but as it *truly* is. Ask any diehard what conservatism is; he'll tell you it's *true* socialism. And the brewer's trade papers: they're full of articles about the beauty of true temperance. Ordinary temperance is just gross refusal to drink; but true temperance, *true* temperance is something much more refined. True temperance is a bottle of claret with each meal and three double whiskies after dinner. . . . What's in a name? " Anthony went on. " The answer is, practically everything, if the name's a good one. Freedom is a marvellous name. That's why you're so anxious to make use of it. You think that if you call imprisonment true freedom, people will be attracted to the prison. And the worst of it is you're quite right " ' (Quoted Stevenson, pp. 214-215).

6 Figuring

6.1 The older and wiser geisha of traditional Japan are said to have lived by the maxim: ' Where figures are deficient, grace and charm must come to the rescue. ' But where figures, in another sense, are more abundant neither grace nor charm can substitute for critical alertness. Consider, for instance, the statement that 70% of the people currently serving sentences in the prisons of Ruritania have served one or more previous terms. Does this mean that Ruritania is afflicted with a recidivism rate of 70%; that is, that 70% of those sentenced for the first time come back for more? And, if so, are we further entitled to infer that, at least in Ruritania, the threat of prison is as a deterrent largely ineffective?

6.2 A moment's thought will show that the correct answer to both questions is ' No '. Take the second first. Suppose that 70% of those sentenced for the first time do come back for more; and let us assume also—perhaps too optimistically—that Ruritania's courts never convict the innocent. Then we certainly may, indeed we must, conclude that most of those sentenced for the first time—70%, to be precise—are in fact not deterred from committing a second offence. This is a sad, bad business: a series of terms in prison is no way to fill a lifetime. But nothing has been said to preclude the possibility that the existing Ruritanian penal arrangements, whatever their faults, are successfully deterring many others who do not, but but for the existence of these arrangements would, offend. The mistake is to overlook that the Ruritanian prison population, precisely because it contains all and only those serving terms of imprisonment, is not a fair sample of the whole adult population of Ruritania.

6.3 Now take the first proposed conclusion. Given that 70% of those currently serving sentences have been in prison before, can we conclude that 70% of those imprisoned for the first time come back for more? No, we cannot. Our information is inadequate to support any such conclusion. If we want to know what proportion of first term convicts eventually graduate into second term convicts, then we shall have to investigate, not the term distribution in the present prison population, but the recidivism pattern revealed by past and future prison records.

6.4 Consideration of this item from the annual returns of the Ruritanian Home Office was intended to show how easy it is to go badly wrong in an important matter by mis-construing even one solitary figure. The lesson should be the more salutary since in this imaginary case we cannot attribute our misinterpretations of the data to the deceitful wiles of someone else, and since we were not entertaining the possibility that the single statistic provided might be itself unreliable. Obviously the possibilities of error may increase with any increase in the quantity and complexity of the statistics available; and obviously too to broach the question of the reliability of the figures offered is to open up a new dimension of difficulty, and maybe duplicity. What often is exaggerated is the importance of the intention to hoodwink other people as a cause of the misinterpretation of reliable figures, as opposed to the compilation of unreliable ones: ' Figures cannot lie, but liars can figure '.

6.5 Certainly a scandalous amount of jiggery-pokery with statistics is executed in the intention, frequently fulfilled, of deceiving other people about what various figures, which are in themselves uncooked, do really prove. But there is also an abundance of self-deception; as well as a deal of error which is not the result of bad faith on anyone's part. Since this book is for those who want to improve their own thinking, it is right to concentrate upon our own intellectual lapses and self-deceits. It would no doubt be more agreeable to pillory such notorious public enemies as dishonest advertisers and demagogic politicians. But better thinking, like charity and over-population, begins at home. Nor should we forget that the demagogue, precisely because he does court popularity, would have no incentive to twisting misinterpretation of the data were there no public eager to applaud shabby manoeuvres in support of comfortable prejudices.

6.6 Another thing which that first example was intended to suggest is that there is in our time no escape from figures. The only choice is whether to misinterpret them or whether to interpret them correctly. It is not only that again and again the best evidence is quantitative, although this is both true and important. It is also that many of the most disputed contemporary issues are themselves essentially quantitative. Questions, for instance, about pollution and about conservation are in practice nearly always questions about how much contamination or how much culling is tolerable. We hear that the (additional) radiation risk consequent upon the commissioning of a new atomic power station is about the same as that arising from the natural background (Taylor, pp. 176 ff). Reassured by the thought that what is natural must be all right, we fail to do the baby sum showing that we are now going to expose ourselves to double the previous—the natural—dose of radiation.

6.7 How complacent we really can afford to be about this must depend on what the original risks were, and the factor by which a doubling of the rad index increases the risks. They too will have to be specified in figures. And then to understand how seriously we ought to be worried, we shall need to make comparisons, which will be equally quantitative, with other more familiar dangers: those—say —of being a casualty in a road accident, or those of birth defects which there is no reason to put down as consequences of radiation exposure. For a mother 20 years old, the chances of having a mongoloid child are about 2 in 10,000; by the time the mother is thirty the chances are up to 1 in a 1,000; at age forty they are 1 in 100; and if the mother is in the later forties the chances are between 1 in 40 and 1 in 25. Or consider the unholy trinity of anencephaly, hydrocephaly and spina bifida. If parents have a child with any of these the chances that the next live birth will be similarly afflicted are about 1 in 20; but the chances that the next pregnancy will end in either another such wretch, or a stillbirth, or a miscarriage, are 1 in 4 (Augenstern, pp. 200-201). For the concerned and the compassionate it is today just not on to accept the irrational, albeit endearing, advice of the Editor of the (pre-Franco) Spanish Anarchist daily: ' Let us have no more of these miserable statistics, which only freeze the brain and paralyze the blood ' (Quoted Brenan, p. 155n.).

6.8 Fortunately what is essential is in no way technical; although it might be a good thing to introduce the notion of statistical significance and the making of Chi-square Tests into the compulsory element in the syllabi of our comprehensive schools. What is needed first and most, and all the time, is an unspecialized critical alertness. This is what we show when we refuse to draw wrong conclusions from the news that 70% of the present prison population of Ruritania have already served one or more previous terms, and when we are not lulled by the comforting word ' natural ' into overlooking that $1 + 1 = 2$.

6.9 Consider discussions about equality and differentials of income and of wealth. Nothing could be more characteristically contemporary, and the subject is essentially quantitative. It will not do, notwithstanding that it is daily done, to compare my weekly wage, or the weekly wage of my members, with his annual salary. To do this is to cheat. It is to misrepresent one of the two terms in the chosen comparison by a factor of 52. (By the way: our vocabulary of pay really does include pairs of expressions with the same descriptive meaning but very different non-descriptive overtones, and where the difference is nothing but a function of the social status of the speaker. When I was President of my local branch of the Association of University Teachers I tried, without success, to persuade the brothers and the sisters to lodge a pay claim, rather than to request as was their wont a salary adjustment: ' A horse sweats, a gentleman perspires, but a lady feels uncomfortably hot '.)

6.10 Again it will not do, notwithstanding that this too is daily done, to compare my take-home pay after tax and all other stoppages with his gross income. This malpractice is especially deceitful in countries with more or less fiercely progressive taxation on incomes. For it is a defining characteristic of any system of progressive taxation—as opposed to proportionate taxation or to poll taxation—that it should diminish the real differences, net of tax, between any extremes of gross income. To leave the impact of progressive taxation out of account must be, therefore, to exaggerate the differences in spendable income net of tax; and the cheat is compounded when, as here, the effect of taxation is neglected in only one of the two terms of the comparison.

6.11 These are strong words, condemning flagrant misdemeanours. Such condemnation is likely to stir up a storm of defenders. ' What do you expect a trades union leader to say?' Well, we ought all to expect (prescriptive) trades union leaders, and everyone else, to present an honest case honestly argued; although I am afraid experience has taught us that this is not always to be expected (descriptive) either of trades union leaders or of many others (5.9).

6.12 ' But you can understand why people think up these twisted comparisons, and cherish them.' Certainly, we all can. But the point was precisely and only to bring out that and how they are twisted, and thereby to make such malpractices fractionally less frequent and less welcome. We all know too why people fiddle their tax returns. But that is not in itself an objection to the endeavours of the Inland

Revenue to make fiddles less common. We have here a fresh species within the genus Subject/Motive Shift (4.3). Let it be simultaneously both christened and pilloried: The But-you-can-understand-why Evasion.

6.13 One response to all such objections carries a moral extending far beyond two particular, and peculiarly flagrant, cheats. It is that, in general, unsound argument and outright misrepresentation tend to discredit the causes they are deployed to serve. If I have to multiply the actual difference by 52 in order to seem to justify my indignation, then the uncommitted critic is likely to conclude that the present situation is nothing near so scandalous as I have been trying to make out. It is also wryly incongruous if in a campaign for social or any other sort of justice I make comparisons which are themselves grossly unfair.

6.14 Consider another kind of one-sided financial comparison. Someone mentions a large sum of money. Do we before responding take care to ask ourselves to what it is relevant in the present context to relate this? Suppose he gives a formidable-sounding amount for the total equity investment of the Church of England. He then comments nastily: ' It is easier for a camel to go through the eye of a needle, than for a rich man to enter the Kingdom of Heaven.' Do we take time out to do a quick calculation of what the present yield on that investment would amount to when divided among 10-15,000 clergy; or could we, if we did, bring ourselves to rate that result as, at current prices and by contemporary British standards, riches?

6.15 We see a headline screaming that some company has run up profits running into millions. Do we, before remarking that that must be where the money goes, ask what these millions amount to as a percentage return on capital employed, or how they compare with the company's wage bill or with its total sales? For these are the comparisons which are relevant if we are thinking of the profits as booty available for a shareout among either the com-

…any's employees or its customers. If our interest really is
…n where the money goes, then we should recall that over
…alf of these profits will be appropriated by the Chancellor
…f the Exchequer, as Corporation Tax; before he proceeds
…o take his second cut of the remainder, as Income Tax on
…he eminently taxable ' unearned ' income of the share-
…olders.

6.16 When we ask the right questions about figures, and
…ctually do relevant rough calculations, we may be sur-
…rised by what we find. We certainly shall be when we turn
…rom questions about income to questions about wealth.
…he distinction is important. It is as easy as it is common
…o think of all wealth as available for spending. If any
…ndividual comes into a bit of money from a legacy, or
…coops the pools, then there is nothing to stop the spending
…f the whole lot in one gigantic spree. But it is a rather
…lifferent story if we are talking about the wealth of a nation;
…nost of which is in form of the physical means of production,
…listribution and exchange. We cannot collectively sell off
…ny of these assets, and spend the proceeds, without thereby
…rejudicing our collective prospects of earning a tolerable
…iving in the future.

6.17 This is another instance of a general truth exempli-
…ied differently in the previous chapter (5.22). We cannot
…ssume, either that what applies to the whole of some group
…aken collectively applies equally to any and every individual
…nember of that group, or the other way about. To take
…nother example: anyone can park here at any time; but it
…loes not follow that everyone can do so simultaneously, or
…ven successively. The crucial practical difference is that,
…while not everyone wants to park here, or at any rate not at
…he same times, almost everyone would like to lay their
…nands on some of someone else's wealth, and to spend it as
…ncome.

6.18 One contributor to a recent Penguin Special writes:
…' the richest 2 per cent of British adults own 75 per cent of all
…private wealth, while the income of the top 1 per cent of

incomes is in sum about the same as that shared out among the poorest third of the population ' (Cockburn and Black burn, p. 165). Before we unleash any indignation against the hated and envied rich we must ask, both where these figures come from, and upon what principles they were compiled. Unless we have some satisfactory answer to the first of these two preliminary questions, we have no guarantee that the figures were not simply invented. In fact the author, although at the time an Assistant Lecturer in Sociology at the London School of Economics, and as such paid to know better, cites no source. Our unease ought to be increased by noticing that in his income comparison he is, apparently, taking no account of the surely considerable impact of progressive income taxation.

6.19 It is important to press the second question too. Users of statistics, unlike users of television sets, need to know something about how the materials were originally gathered and later cobbled together. Without this information we may easily draw quite unwarranted conclusions. Were these particular figures, for instance, derived from Estate Duty returns; and, if so, what if any allowance was made for the fact that most estates are too small to attract this tax? The sum of all these individually small estates could very well be, relative to the sum of all the biggest estates, large; and whether or not it in fact is, is a main part of precisely the point which is here in question. Again, what allowance, if any, was made for pension rights and life assurance policies? The former escape the Estate Duty net completely; whereas the latter, in so far as they are caught, are bound to come up tagged with a higher value than any which could have been imputed earlier. Yet pension rights and life assurance policies, usually backed by investments held in the name of some corporation, must during life constitute a significant proportion of the wealth, such as it may be, of most middle class people.

6.20 *The Economist* recently made and published some calculations about the distribution of wealth in an imaginary country, not Ruritania this time but Egalitaria. These are relevant to us in three ways. First, they constitute another and even more impressive example showing how unexpected may be the results revealed by still not intolerably difficult calculations. Second, they underline the curiously neglected importance of the fact that people progress through life cycles from infancy to old age; and that we can scarcely expect, in either sense of ' expect ' (5.9), to be in the same financial situation at every stage in our particular cycle. Third, they spell out some of the unnoticed implications of one set of egalitarian ideals. They thus provide those who nourish such ideals with a very necessary frame of reference, enabling them to assess more accurately how far and where the actual situation falls short of their own aspirations.

6.21 All Egalitarians are educated publicly up to the age of 21, with no opportunity at this stage to earn enough to save. All men then work for the same wage till 65, when they retire on full pay: women work for only twenty non-childbearing years, but in those years get equal pay with the same pension rights. Inheritance is forbidden. But all earners and all pensioners save exactly 10 per cent of their incomes, which savings are invested in state bonds yielding 10 per cent compound. This rather high rate perhaps compensates for the absolute embargo on all capital appreciation. The roundness of the figure also simplifies the arithmetic, as do two further stipulations: that the net reproduction rate has been unity for the past 85 years; and that everyone dies on their 85th birthday. So how much of the privately owned wealth in Egalitaria is owned by the richest 10 per cent of the whole population? ' The answer seems to be that the wealthiest 10 per cent of Egalitarians (who by definition are all the men aged from 68 to 84 inclusive) must now own about 74 per cent of the privately owned wealth . . . ' (26/v/73, pp. 16-18).

6.22 A main theme of the present chapter is that what is needed first and most, and all the time, is not technical

expertise in statistics but an unspecialized critical alertness
The way to acquire this, or any, disposition is by practice
and to underline that it is a lay rather than an exper
business it may help if some of the practice is on specialists
Edmund Leach is one of the leaders in British anthropology
In the third of his Reith Lectures, as printed in *Th
Listener*, he said: ' Admittedly the statistics show a numerical
increase in the incidence of crime. But this is a measure of
police efficiency, not of the moral state of the nation.
Crimes are created by Parliament: it needs a policeman to
make a criminal ' (30/xi/67).

6.23 Certainly it is true, true by definition, that, in the
legal sense of the term ' crime ', what is and is not here and
now a crime is determined by what the law says: private
homosexual relations between consenting males over 21
years old ceased to be criminal in England on the day when
the corresponding change in the law took effect. So there is
a sense in which our criminals are made (not by our
policemen but) by our Parliament. But from this purely
verbal premise we are not entitled to deduce the massively
substantial conclusion that either our legal system or our
policemen are in fact so counter-productive as actually to
induce people to commit such acts as happen to be pro-
scribed as criminal.

6.24 This is one more occasion where we need the
fundamental distinction between propositions which are
analytic and logically necessary, and propositions which are
synthetic and contingent (3.4-3.5). Another similarly
fallacious move is currently a commonplace in Third World
First circles. It is no doubt necessarily true that the rich
make the poor, and the other way about; in the sense that
such correlative terms as ' rich ' and ' poor ' depend for
their meaning upon the possibility of mutual contrast. But
from this alone we cannot legitimately infer—what I am
not now asserting or denying—that rich men, and rich
countries, always and everywhere grind the faces and pick
the pockets of the poor.

6.25 Having disposed of Leach's third statement instructively, we can tackle now the more immediately relevant first two. Sophistication is here going to Leach's head. The sophistication is to be aware, in general, that even the most honestly and competently compiled figures may not mean what they seem to mean, and, in particular, that they may sometimes tell us more about the method of compilation than about their supposed subject. Yet to assume that this must always be so, or to assume without particular reason that it is so in some particular case, is to be infatuated with your own insight. However: there but for the grace of God go we. For when we first become acquainted with some fallacy it can be almost irresistibly tempting, as students of the present book need to recognize, to identify arguments as tokens of this fallacy type when what we really have is, either a token of some other type, or even no fallacy at all (1.46).

6.26 The insight by which Leach has allowed himself to be carried away is important, and it provides one of the reasons why we have to ask both where figures come from, and upon what principles they were compiled. Suppose that our eye was caught, in the old pre-Wolfenden days, by the exhilarating headline: 'Wave of Buggery in Bootle'. It may be that this was touched off by a spectacular rise in the number of successful prosecutions, and it may be that this figure has been misinterpreted. For the police do have, as they have to have, some discretion about what is to be investigated most strenuously; and perhaps a new Chief Constable had recently been appointed, who had a thing about male deviants. If so, then maybe that headline ought to have been, even better: 'New Broom Sweeps Bootle's Buggers'.

6.27 Legend tells too of a feminist research student who carried out a wide and weary psychometric programme, coming up eventually with the gratifying finding that the average IQ of men is equal to the average IQ of women. It must have been hard to tell her that one of the specifications for the particular battery of tests which she was applying was that those chosen must yield this result.

6.28 It is not, however, legend but history that Karl Pearson, a leading statistician and philosopher of science, once published a paper on ' The Scientific Aspect of Monte Carlo Roulette ' (Pearson, Vol. I, pp. 42-62). His conclusion was, with appropriate qualifications, that ' roulette as played at Monte Carlo is not a game of chance'. This result was derived from his analysis of the outcomes of every spin made, day by day and week by week, for many months; which outcomes were all recorded, for the benefit of seekers for systems, in a paper called *Le Monaco*. It later emerged that the reporter assigned to this dispiriting task had preferred to sit out his days at a nearby café rather than to stand on tiptoe peering at the tables in the casino. So Pearson's statistical analysis was really not evidence of some hidden, and perhaps exploitable, physical regularities in those spinning wheels; but evidence for the wider psychological fact that if we try to think up, straight out of our own heads, a genuinely random series, we shall not succeed.

6.29 It is, therefore, vital to be alert to the possibilities, that sometimes figures may simply have been conjured out of the air, or that what they really reveal is if anything something about the conditions of their compilation rather than something about what they seem or purport to record. But these are only occasionally realized possibilities and not, even in social studies, general necessities. No recognition of the dangers of mendacity, of ingenuous error, and of self-deceit, ought to blind us to the indispensable necessity of quantitative evidence and quantitative thinking. If we are committed to the improvement of our understandings we must categorically reject as obscurantist the slogan: ' There are lies, damned lies, and statistics.'

6.30 Consider yet another illustration, the last in the present chapter. Recently the Librarian of the City

University wrote to *The Times Higher Education Supplement* about ' the astronomical increase in the prices of periodicals published by the societies ', and, after remarking the consequent difficulties for academic libraries, commented: ' one has not failed to notice the increased profits culled by publishers. The net income of what I take to be a typical periodical publishing house increased by 5.5 per cent in 1972 as compared with 1971. . . . Is profit growth the only motive? ' (16/xi/73).

6.31 Let us waive, perhaps with regret, questions about the much abused profit motive. (Why, for instance, is our economic psychology never enriched by references to the wages motive, the salary motive, the fixed interest motive, or even the best-buy motive?) The point for us now is that the City Librarian is so rightly distressed by the impact of inflation upon his own charge that he fails to take account in his subsequent argument of its impact upon others. Since, however you measure it, the rate of inflation between whatever are the two precise periods in question was certainly more than 5.5 per cent, what is here called a 5.5 per cent growth in profits must have been, in real terms, a decline. This conclusion is a necessary consequence of the consistent application of the principle, now universally recognized, that anyone whose earnings, or pension, has not kept pace with inflation has suffered, in real terms, a cut.

6.32 Another manifest but infrequently mentioned consequence is that, in so far as no discount is made covering actual inflation, any Capital Gains Tax is necessarily a Wealth Tax; albeit a Wealth Tax exacted not regularly each year on all exposed possessions, but irregularly on whatever sales of these have been made in any particular tax year.

6.33 Yet another equally undeniable consequence, and one with formidable and still far too narrowly appreciated economic consequences, is that until and unless your systems of public and private accounting make due allowances for inflation all your balance sheets must present a distorted picture. Since, as all actual and would-be home-owners know, the current money value of what estate agents call property tends to rise faster even than the general inflation level; such property, left on the books at its cost in the money of the year when it was bought, will tend as the years pass to be increasingly undervalued. In the sixties wide-awake operators made takeover bids for many companies whose sleepy managements had failed to realize this, and who were, in consequence, not making the most effective use of these appreciating assets.

6.34 But of course most of the assets of most companies are various sorts of equipment and stocks of materials. The former wear out and become obsolete. So they have in a few years to be replaced. If your annual depreciation allowance, set aside to cover the costs of eventual replacement, is too small to meet those costs, then you are, in the years when you make such inadequate allowances, running down your capital. In so far as these depreciation allowances are what would be the appropriate fraction of the original cost in pounds that stayed put, they are bound to be too small in a period of inflation; and the more rapid the inflation the more inadequate the allowance determined by any such formula. Since profits are the residue after the subtraction, among other things, of depreciation, any inadequacy in the allowances for depreciation must correspondingly exaggerate the figures for profits, if any.

6.35 The most formidable economic consequence of refusing to adjust our systems of public and private accounting to the realities of the age of inflation is that today many, perhaps most, of those industrial firms which, given present bookkeeping practices, appear to be making profits, are really only achieving this appearance by, in real terms, running down their total assets. If and in so far as this is so any dividends which are being paid are not being paid out of current real earnings, because in real terms there is only a loss. The 52% Corporation Tax which is levied on these notionally positive but really negative current earnings, and then spent by the Chancellor as income, must be, again in real terms, a levy on the nation's industrial capital. Yet this industrial capital must be maintained, and increased, if we are, as a nation, to earn as good, or a better living in the future (6.16).

6.36 The truth of the important economic and fiscal conclusions suggested in the previous two or three paragraphs is, however, for the moment none of our business. The concern of the present book is with the quality of argument in all areas, and not with the truth of particular conclusions in any. The examples drawn from this and many other fields have been cited primarily to illustrate points which are, in a broad sense, logical. But often, as here, they have been chosen also to suggest, as less interesting and possibly less distracting illustrations might not, that logical points matter. The crucial suggestion in this particular case is that the financial, political and economic consequences of failing thoroughly and consistently to appreciate the slipperiness of the key term can be very serious indeed, not to say catastrophic.

6.37 The moral for us is that our money measures have become Systematically Ambiguous. It was Aristotle who first drew attention to certain terms which have different meanings when applied to different sorts of things, yet the same meanings when applied to the same things. Our money measures now have a new variety of such Systematic Ambiguity, Systematic Ambiguity in the time dimension. A pound is a pound is a pound anywhere at any one time. But a pound (1974) is not, but is less than, a pound (1973). We can only pray, if not hope, that we shall get off the Gadarene slope before months, or even days, have to be intruded into those essential parentheses.

7 A Chapter of Errors

7.1 It is a pity that Chapter Six had to stop when it did.
Nothing but good could have come from examining more,
and more various, ways in which figures can be misinterpreted. It is equally a pity that the present chapter will
again be far too short. In both cases the examination of
more, and more various, examples is the practice which can
scarcely fail to improve practice. Yet the first thing is to
appreciate that what is needed most, and all the time, is an
unspecialized critical alertness. We have to be always
ready to discern a fault in argument in any field; and most
especially when that argument is our own, or seems to
support some conclusion which we ourselves cherish. Both
where figures are involved, and more generally where they
are not, the faithful treatment of a few examples should be
sufficient to emphasize the need; and to indicate the kind of
discipline required in order to do better.

7.2 There is a tale, probably apocryphal, told of that
notoriously merry monarch Charles II. There was a dinner
to commemorate the foundation of the Royal Society. At
the end of the evening, ' with that peculiar gravity of
countenance which he usually wore on such occasions ', he
put a challenge to the Fellows. ' " Suppose two pails of
water were fixed in two different scales that were equally
poised, and which weighed equally alike, and that two live
bream, or small fish, were put into either of these pails," he
wanted to know the reason why that pail, with such addi-
tions, should not weigh more than the other pail which
stood against it.' Many suggested possible explanations,
and argued for their own suggestions with more or less

vigour. But at last one who perhaps remembered that the motto of that great society is ' Nullius in verba ' (Take no man's word for it!), denied the assumption: ' It would weigh more '. The King was delighted: ' " Odds fish, brother, you are in the right " ' (D'Israeli, p. 341).

7.3 The king's move was an instance of what has traditionally been called The Fallacy of Many Questions. The stock example is the question ' When did you stop beating your wife? '; put to a man who either is not married at all, or else has not started, or perhaps has started but not stopped, beating his wife. It is not clear that what is wrong here is, in the strictest sense, a fallacy (1.48). But obviously it is wrong to build false assumptions into a question, and to give answers which accept such assumptions. The crux is noticed in Chapter Six of *Alice's Adventures in Wonderland*: ' " How am I to get in? " asked Alice again, in a louder tone. " Are you to get in at all? " said the Footman. " That's the first question, you know." '

7.4 Unfortunately these, like all the best text-book illustrations, make it so clear what is going wrong that it may become hard to believe that anyone could make seriously and ingenuously such a mistake. But a crucial and contestable assumption may be very hard indeed to unearth when it is concealed in a longer and more complicated statement. Even when it is embodied in a single and lucid sentence we may well miss it. Take Hugh Gaitskell's words in the British budget debates of 1959: ' the budget involved giving away not far short of £400 million. Any Chancellor who could give this away could be described as lucky.' Or take the sentence, found in what Professor Ninian Smart thinks of as a neutralist treatment of *The Religious Experience of Mankind*: ' We have records of the inaugural visions of some of the Old Testament prophets, of the experiences which taught them something profoundly important about God, and that spurred them on to teach men in his name ' (Smart, p. 22).

7.5 When Hugh Gaitskell spoke of the Chancellor

' giving away not far short of £400 million' (1959), the reference was to tax cuts. His description has since become a commonplace: which is more than can be said, unfortunately, of actual tax cuts. Yet this description apparently assumes that all property is the creature of the state; and hence that whatever is not taken from us, must be given to us, by the state. This is a view in the Hobbist tradition; and it is thus at the opposite pole to the Lockean, holding that the reason why there must be no taxation without representation is that my consent, whether given directly or through representatives, is prerequisite to any transfer of my property, whether into the public purse or anywhere else. Gaitskell himself may very well not have appreciated the full significance of this Hobbist assumption. Certainly it would have been rejected by everyone on the other side of the House, as well as by many on his own.

7.6 The assumption concealed in the second passage is built into the phrase ' the experiences which taught them something profoundly important about God '. By writing these words the author implicitly, and persumably without completely grasping this implication, claims religious knowledge in the strong as well as the weak sense (5.12). For there is a decisive difference: between, on the one hand, affirming that someone enjoyed or suffered such and such experiences, which experiences led him to believe this, that or the other; and, on the other hand, while still affirming that the experiences occurred, also conceding that those consequent beliefs were in fact true (Flew (2), Chapter 6). If the beliefs in question are religious, and if you are supposed to be writing a detached essay on comparative religion, then you have no business either to assert or to imply any propositions of the latter sort; not even if you do believe, and maybe rightly, that they happen to be true.

7.7 The Fallacy of Many Questions is not strictly a fallacy, although the importance for thinking of noticing what is presupposed not only by questions but also by statements cannot be over-emphasized. The Genetic Fallacy really is a fallacy, and it consists in arguing that the antecedents of something must be the same as their fulfilment. It would be committed by anyone who argued, presumably in the context of the abortion debate, that a foetus, even from the moment of conception, must really be, because it is going to become, a person. But the fallacy is more usually exemplified in equations moving in the opposite direction, from the actual or supposed antecedent to the developed whatever it may be.

7.8 Consider a recent best-seller, hailed by its delighted publishers as ' a wildly successful book '. My copy of *The Naked Ape* reports that the reviewer for *The Times Educational Supplement* described it as ' brilliantly effective, cogently argued, very readable '. Marshall McLuhan agreed: ' As with the title, the entire book is full of fresh perception.' Explaining that wildly successful title the author says: ' There are one hundred and ninety three living species of monkeys and apes. One hundred and ninety two of them are covered with hair. The exception is a naked ape, self-named homo sapiens ' (Morris (1), p. 9).

7.9 Since McLuhan went out of his way to commend the title, it is just worth pointing out that the opposites of ' naked ' and ' covered with hair' are, respectively, ' clothed ' and ' hairless '. So any ' fresh perception ' here has resulted in a misdescription. It is, however, a misdescription which suits the author's purpose. This is metaphorically to strip man and, as he was to express it in a sequel, to reveal ' a human animal, a primitive tribal hunter, masquerading as a civilized, super-tribal citizen ' (Morris (2), p. 248). Certainly it can be salutary to be reminded that, whatever else we are or may become, we remain animals: ' Even a space ape must urinate '(Morris (1), p. 21). And certainly it is useful to insist that as animals we have inescapable problems generated by our fertility. But simply to identify us with our nearest ancestors on the evolutionary family tree is an altogether different thing.

It is this which, again and again, Morris does: ' Behind the façade of modern city life there is the same old naked ape. Only the names have been changed: for " hunting " read " working ", for " hunting grounds " read " place of business ", for " home base " read " house ", for " pair bond " read " marriage ", for " mate " read " wife ", and so on ' (*Ibid.* p. 74). Once more: ' When you put your name on a door, or hang a painting on a wall, you are, in dog or wolf terms, for example, simply cocking your leg on them and leaving your personal mark there ' (*Ibid.* p. 161).

7.10 The nerve of the argument, and it is an argument which comes up all over the place, is that if this evolved from that, then this must always be that; or at least, it must always be really or essentially that. Yet a moment's thought shows that this argument is absurd. For to say that this evolved from that implies that this is different from that, and not the same. It is, therefore, peculiarly preposterous to offer as the fruit of evolutionary insight a systematic development of the thesis that we are what our ancestors were. Oaks are not, though they grow from, acorns; and—for better or for worse—civilized people are not, though they evolved from, apes.

7.11 It is, furthermore, also egregiously preposterous to present as biological understanding a wholesale depreciation of environment as opposed to heredity, of what is learnt as opposed to what is instinctual. Certainly the distances between man and the brutes, and between city folk and primitive people, are in this perverse perspective narrowed. But it conceals what is the most distinctive and powerful peculiarity of our species, as compared with all others. This concealment is a very queer thing to parade as a zoologist's revelation.

7.12 The peculiarity is our comparatively enormous capacity for learning, and the length of the period of the upbringing of children. This learning capacity, and its main trophy and instrument language, provides our most favoured species with an excellent substitute for the inheritance of acquired characteristics. I am especially pleased to thank Julian Huxley for that last point. For I found it in his *Essays of a Biologist*, which is a fine work of popularization. It demonstrates that it is not necessary in order to achieve popularity to sensationalize your subject by standing it on its head.

7.13 The tempation to commit the Genetic Fallacy is greater when the evolution is thought to have been smoothly continuous. For it is then reinforced by the different temptations of The Logically-black-is-white Slide. This unsound form of argument is extremely popular, although it is rarely made fully explicit. It runs roughly as follows: ' The difference we are dealing with is a difference of degree. Since with such a difference there can be no natural break at which a sharp line of division is, as it were, already drawn, there can be no logical stopping place on any journey from one extreme to the other. The difference cannot, therefore, be one of either kind or principle. So it must really be either non-existent or, at best, unimportant.'

7.14 The first thing is to get a little clearer about differences of degree. Let us say that a difference of degree between two extremes is one such that there is, or could be, a series of actual cases, or of theoretically possible cases, stretching between one of these extremes and the other, and with the amount of difference between each member of the series and the next vanishingly small. It becomes obvious that in any such series ' there can be no natural break at which a sharp line of division is, as it were, already drawn '.

7.15 But it should also be obvious that differences which are both large and in this sense differences of degree can be of the last importance. For the differences between age and youth, between riches and poverty, between sanity and insanity, between a free society and one in which everything which is not forbidden is compulsory, are all paradigm cases of large differences which are differences of degree in the sense just now explicated. It must, therefore, be as wrong as it surely is common to move as if no move had been made, and hence without any particular justification: from saying that this is a difference of degree; to saying that this is a mere difference of degree.

7.16 What starts people down on the disastrous Logically-black-is-white Slide is the observation that some difference is a difference of degree, and that ' there can be no natural break at which a sharp line of division is, as it were, already drawn.' This is, as we have seen (7.14), a defining characteristic of differences of degree. What is wrong is to assume that it warrants the conclusion that even the largest of such differences ' must really be either non-existent or, at best, unimportant '. It is possible to disentangle, a little artificially, three separate strands in this error.

7.17 First, is the notion that, if a distance is one which is or could be travelled in a series of very short steps, then it cannot even in sum amount to much: a muckle cannot be made of no matter how many mickles. This wretched misconception has a Classical label. It is called either The Sorites (Greek, pronounced *So-wry-tees*) or The Heaper: ' Since a single grain of sand does not make a heap, and since adding one more is at no stage enough to convert what we have into a heap, there cannot really be heaps.'

7.18 Second, is the idea that we cannot properly distinguish between this and that, much less insist that the difference is desperately important, unless we are able to draw a sharp line between one and the other. Of course the lack of such a line may sometimes be very inconvenient. Still it does not even begin to show that when the items distinguished are well clear of the undemarcated no man's land, we cannot or should not make such distinctions. As Edmund Burke once said, with his usual good sense, ' though no man can draw a stroke between the confines of night and day, still light and darkness are on the whole tolerably distinguishable.'

7.19 Third, is the assumption that, where there is no

bvious natural break, and hence where any line drawn
ust be artificially drawn; there, both our drawing of any
ne, and our decision to draw that line at one particular
oint rather than another, must be illogical. These are the
ost important misconceptions. They nevertheless are
isconceptions.

7.20 Here recall what was said earlier about the reasons
hy logic sometimes gets a bad press (1.31-1.38). The
cond of these was ' that it is confused with various things
at have nothing to do with it . . . there certainly is nothing
gical, in our primary sense, about impractical practices or
nworkable institutions ' (1.35). One source of trouble is
at the logical may be contrasted either with the illogical
r with the neither logical nor illogical. The rational may
milarly be contrasted with either the irrational or the
either rational nor irrational. The correct interpretation
f the words ' logical ' and ' rational ' must depend on what
, in the context, the correct contrast. It is, therefore, a
istake to take it that what is neither logical nor illogical,
either rational nor irrational—we might speak of the non-
gical, or of the non-rational—has to be by the same token
logical, or irrational. It is also a dangerous mistake. It
ncourages those, who, as all of us do, value some things
hich are non-logical, or non-rational, invalidly to infer
at these humanly indispensable values militate against, or
ven preclude, logic and rationality. Certainly there is a
lace in life for passion, for compassion, and for commit-
ent. For the illogical and the irrational there is, or ought
o be, none.

7.21 Descending from the rather abstract and theoretical
evel of the previous few paragraphs, and considering some
f the paradigm cases of differences of degree already
entioned, it is easy to see that we often have the most
xcellent reasons for drawing a line. Suppose we want to
ount a round-the-clock operation. Then we shall need to
ix some pretty precise point in the diurnal cycle at which
he day watch is to take over from the night watch, or the

day shift from the night shift. Suppose we want to introduc
a Wealth Tax. Then we shall need to specify exactly wha
total of what sorts of resources is to count as, ' within th
meaning of the Act ', wealth. If certain legal rights an
duties are to be attached to adult status, then it become
practically essential to determine some moment in people'
lives when they become legally adult. And so on. There i
absolutely nothing illogical about being led by such practical considerations to draw sharp lines, lines which necessaril
are, in the sense explained (7.14), artificial. And, furthermore, since the choice to do this is reasoned it is diametrically wrong to berate it as arbitrary.

7.22 There may also be, and again there often are, good
reasons why the line to be drawn should be drawn throug
this particular point rather than that, or at least good
reasons why it should be drawn through one of the points
of this particular sort rather than through any of the rest
For instance: given that we have some standard units fo
measuring the difference in question, then everyone wil
find it easier, more natural as opposed to forced, if any lin
which is to be socially important—not to say socially
divisive—is drawn through some point corresponding to a
whole and preferably a round number of these standard
units. Even though the choice between 18 or 19 or 20 or 21
as the age of majority may have to be unreasoned, and
hence arbitrary, the decisions to fix some age of majority,
and to fix it so that the line is drawn at a birthday rather
than at any other day in the year, are by no means unreasoned, and hence not arbitrary.

7.23 Whatever lines we draw across any differences of
degree are bound to generate paradoxes of the little more
and how much it is, and of the little less and how little:
' If only Jack had been born an hour sooner he would not
have been liable to call-up, and he might have been alive
today; if only Jill had not managed to have that little bit
less, she would have been clobbered by the Wealth Tax.'
The argument may then be offered that so much ought not

hinge upon these so littles. To this the only but sufficient general reply is to insist that, just because so many humanly vital differences are differences of degree, and just because we do for the best of reasons have to draw artificial lines of division across these continuous differences, the occurrence of some such paradoxes, with all the consequent heartburn or self-congratulation, is an inescapable part of the human condition.

7.24 What we must on no account do, once such lines have been established, is to pretend that the particular little more or little less is all that is now in jeopardy. For the line laid down was a line of policy, and in that sense of principle. So what is now in question is not only that particular little more or little less, but also, for better or for worse, the general policy or principle which any breach must challenge. Thus there were in the summer of 1939 a few who complained that we were to engage in a war over Danzig. This was totally wrong. Since a policy had been adopted by drawing and holding a line, what was at stake was not only Danzig, but also that whole principle of resistance to Hitlerite expansion.

7.25 So far none of our illustrations has been drawn from advertising. So much is in some circles urged or assumed about the alleged evils of this activity that the absence of such illustrations may strike many readers as curious. But it becomes less surprising when we consider how small a proportion of total advertising output is argumentative prose. Certainly when I checked through the advertisement pages of the various newspapers which come into our house each week I found that most of the column inches of both large and small advertisements just gave information about what was ' For sale ' or ' Wanted '. Of course some of this information may have been, or might have been, false. But any problems here are not problems for us now but for those who drafted, passed, and may one day amend the Trades Descriptions Acts.

7.26 Naturally such information will usually, when there are alternatives, be presented in a way favourable to the advertiser's interests, in the broadest sense of the word 'interests'. There is a difference, which is not a difference in the information given about the beer, between saying that the mug is already half empty, and saying that it is still half full. My own two favourite examples of such accentuation of the positive are, perhaps fittingly, both North American. In response to distress about the environmental effects of phosphates one wideawake firm in the USA spread the exultant sales message that its domestic detergent was '98.8 per cent phosphate free'. The Atomic Energy Commission in the same country has become similarly careful to promote what used to be called Hazard Analyses as, more encouragingly, Safety Analyses (Taylor, p. 185; Taylor is, I think, too harsh in scoring this sort of thing as deceitful).

7.27 More serious, and of more obvious concern to us here, are certain other ways of putting a better face on percentages and similar quantitative comparisons. The first and most fundamental question - to ask about all percentages is 'Percentage of what?' As we emphasized in Chapter Six, 70% of those now in Ruritania's gaols is not at all the same thing as 70% of all Ruritania's first offenders (6.1-6.3). The news that profits in retailing have increased or are to be by law reduced by 10% tells us nothing in particular until we know whether this is a percentage of mark up over wholesale prices or of gross profit on capital employed. It is the former which directly determines prices in the shops. But that is very different from, and is almost bound to be enormously larger than, the latter; since it is from the former that the retailer has to meet all his costs—premises, wages, equipment, and so on.

7.28 Suppose you boast that the New Splodge contains 50% more of some gorgeous ingredient. Then your boast is as near as makes no matter completely empty if no one knows, and you are not telling, how much there was in before. Such no doubt true statements can be doubly misleading thanks to certain very simple mathematical properties of percentages; the birth of your second child constituted a 100% increase in your family, whereas if you were to have a fifth that one would represent only a 25% addition to your previous four; if you were to suffer a 50% cut in anything you would then need a 100% increase to get back to where you were before.

7.29 Other simple mathematical properties are important in the pictorial representation of data comparisons. The area of any plane figure, if the shape is held constant, increases much faster than any of its dimensions. In the simplest case, that of the square, the doubling of the length of the sides quadruples the area. It will not do, therefore, to represent the gratifying rise in the sales of your electric toasters by a series of pictures of toasters, one for each year, and in which the sales figures correlate with the height or the width of the sketches. For the increases in the areas of these sketches must then be far greater than the sales increases which they purport to illustrate. There is a similar disparity between the increase in the area of a square and the growth in the volume of the corresponding cube. It may be helpful as a mnemonic to remark that it was part of the genius of Isambard Kingdom Brunel to see the relevance of this to the problem of building a steamship capable of carrying enough fuel for an ocean crossing: 'This is that whereas the carrying capacity of a hull increases as the cube of its dimensions, its resistance, or in other words the power required to drive it through the water, only increases as the square of those dimensions' (Rolt, p. 249).

7.30 Mistakes and misdemeanours exploiting these and other simple mathematical properties certainly are to be found in advertisements; as well as, perhaps even more commonly, elsewhere. Then again you might, I suppose, deem all the attempts made by salesmen to build up agreeable associations for their products to be arguments, and i so certainly bad arguments, for buying these products But to do this surely is to stretch the word 'argument' a bi too far to be helpful?

7.31 It also involves what looks to me like a misunder standing of both the aims and the effects of such campaigns Only someone with an excessively contemptuous view o British stout drinkers and British margarine buyers would believe that they are being manipulated by the not ver hidden persuasion of Guinness or Summer County poster into buying products which they do not want and which but for all those irrelevant cartoons and irrelevant picture of unspoilt countryside, they would not prefer to the available competition. For the advertiser the aim is, pre sumably, to generate goodwill and to keep the name of the product in the public mind. And if, all other things being equal, I choose Pirelli tyres because I have enjoyed their calendars, is that a bad or a silly reason?

7.32 I can, however, offer from my philosophical scrap book one prize specimen of a reason which is both bad and silly. It comes from a Silvikrin advertisement published in the since defunct *Picture Post*. Under the heading 'Can baldness be postponed?' we read: 'Faced with fast-falling hair most people make some attempt to delay the evil day when baldness can no longer be denied. Some try to disguise the fact with long forelocks and other subterfuges. But the wise and knowledgeable face up to the fact that their hair is dying from "natural causes" and that a *natural treatment* is the only hope of saving the situation' (9/ix/50: italics original). He who drives fat oxen must himself be fat!

7.33 A more interesting example, more worthy of our thinking steel, comes not from an advertisement but from two critics of advertising: it will thus also do something to restore the balance. The authors charge: 'Advertising contributes to cultural stratification within our society.' This is urged on the grounds that all the advertising for one particular product 'will be adapted in style, appeal and

ophistication to the appropriate audience' (Hall and
Whannel, p. 315). But, first, it really will not do simply to
assume that this adaptation is the cause and not the consequence of there being specialist journals catering for
specialist publics. Nor, second, must we take it as obvious
that our ideal should be one uniform and homogeneous
audience, with no minority publics served by specialist
minority journals.

7.34 The authors next proceed—and here we come to the
interesting part of their argument—to deploy examples
which are said to ' show how journals which are generally
considered to be media carrying news, features, opinion,
information, or devoted to specialized hobbies, become
transformed into means of selling products and penetrating
markets' (*Ibid.*, p. 321). Yet none of the examples are
examples of the actual transformation of anything. They
are instead quotations from the advertising trade press,
making claims about the sort of public to be reached by
advertising in this journal or in that. It is almost embarrassing to have to point out that the *Angling Times* does not
suddenly cease to be devoted to the specialized hobby of
angling just because you as a merchandiser see it as your
way to: ' Get your hooks into the angling market ' (Quoted,
Ibid., p. 315).

7.35 Hall and Whannel have thus provided another, and
bizarre, token of a type exemplified in an earlier chapter
(4.16-4.17). A statement of the particular limitations of
some professional interest, or a statement that someone sees
something only as a such and such, is misconstrued as
warranting the conclusion that what that profession is
professionally interested in, or what that individual person
sees it as, is all that there is there. In the present second
specimen we have a further twist. The professional interests
which lead the advertisers or their business clients to see the
Angling Times only as a means of getting their hooks into the
angling market are themselves seen as magically transform-
ing that journal. Yet from beginning to end no reason has

been given for saying that either the advertisers or their business clients have had any influence on any of the journals in question. Indeed had anyone had such influence as really to make it no longer one of the 'media carrying news, features, opinion, information . . . devoted to specialized hobbies', then advertising in the *Angling Times* could no longer serve the purposes of those fishers of sales who want to get their hooks into the angling market.

7.36 I cannot end without sharing another commercial item from my philosophical scrapbook. Let it serve as a final cheerful reinforcement for what was said earlier about the crucial importance of contradiction and non-contradiction (1.16-1.23). This item comes from a circular letter from Dover Publications, Inc., a New York firm with a very useful line of reprints of classical texts in Philosophy, Science, and—wait for it!—Logic. It reads: ' We shall enjoy hearing from you, and for your convenience we are enclosing a business reply envelope which requires no postage (due to US postal regulations reply envelopes cannot be used from foreign addresses, and therefore none is enclosed).'

8 The Final Foreword

8.1 In so far as any book of the present sort is successful, its end must be a new beginning. Yet for many the effect of appreciating some of the variety of faults committed may be not encouragement but despair. It surely is impossible to avoid all these and other mistakes, and to get everything right? No doubt it is. But it is as wrong here as it is everywhere else to argue that if I cannot do everything, then I cannot, and am not obliged to, do anything. It is, for instance, intellectually but not only intellectually shabby to argue that, because I cannot contribute either money or time and effort to all the causes which may make demands on me, therefore I cannot and need not contribute to any. It is also an error, albeit one committed by the great Immanuel Kant himself, to contend that striving after perfection must presuppose a commitment to the belief that actual perfection can and perhaps will be achieved.

8.2 Perhaps the shortness of the book will reduce such temptations to discouragement: the wise teacher keeps his reading lists short partly lest his students should conclude that, because they cannot read everything, they do not need to read anything. But the main thing to stress again now is that the challenge to think better is a challenge to our integrity.

8.3 In Chapter One I tried to bring out how the notion of valid deductive argument is and has to be defined in terms of contradiction and of non-contradiction (1.1-1.10). No one who has any concern with what is or is not true can afford to be unmoved by the threat or still more by the actuality of self-contradiction (1.16-1.23). In Chapter Three

Thinking about Thinking

I displayed some of the possibilities of self-deceit through covert shifts between substantial and tautological interpretations of the same forms of words (3.1-3.10). I then went on in the same chapter to link this with the Popperian thesis that a forthright concern for truth demands an emphasis upon the possibilities of falsification, and a permanent critical openness towards their realization (3.13-3.20). In Chapter Four my main concern was that any insights which we have into the reasons why people may be inclined to hold or to utter certain propositions (whether these reasons are motives or causes), and any insights which we may gain from the future advancement of psychology and sociology, should be harnessed to the improvement of our thinking, and not misemployed to distract attention from questions about the reasons (grounds) for holding that these propositions are true or, as the case may be, false (4.3-4.20: for a more thorough examination of these different sorts of reason see Flew (1), and the other contributions discussed therein). In Chapter Five I urged that to play Humpty Dumpty with the established meanings of words is to act in bad faith (5.21-5.25).

8.4 It is time at least to suggest a wider connection between rationality in general and personal integrity. This suggestion is the more important and the more timely because it is nowadays fashionable to disdain the former, and in particular the exacting standards of science, in favour of a supposedly incompatible ideal of sincerity and personal relationships. This is a preposterous antithesis, since sincerity and integrity require what is being in their name rejected.

8.5 Back in Chapter One I wrote: ' To say that someone knows something is to say more than that he claims to know it, or that he believes it most strongly. It is to say also, both that it is true, and that he is in a position to know. So neither the sincerity of his conviction nor the ingenuousness of his utterance guarantees that he knew ' (1.54). Someone may be absolutely sincere and ingenuous in claiming to know, and yet nevertheless turn out to have been mistaken. That this can and does happen is both a philosophical and an everyday commonplace.

8.6 What is not quite so often remarked is that to the extent that I make claims to knowledge without ensuring that I am indeed in a position to know, I must prejudice my claims both to sincerity and to ingenuousness. It is just not honest for me to pretend to know the winners of tomorrow's greyhound races when I am not directly or indirectly acquainted with either the form of the dogs or the plans of the dopers. Nor will my dishonesty be diminished— though the consequent damage will be—if my predictions happen to be fulfilled. Notwithstanding that law cannot be equated with morals, nor belief with knowledge, it is to the point to say that by the Perjury Act of 1911 it is perjury to swear what we believe to be false, regardless of whether what we swore to happened to be true (Stephen, Vol. IV, p. 148).

8.7 Being in a position to know is not always, or even most often, a matter of being able to deduce what is known from premises also known. Clearly it could not be. For on this assumption knowledge would be impossible, since it would require the completion of an infinite series of deductions from premises all of which would have first to be deduced from others, in turn first deduced from others, and so on. Sometimes we know without inference, as when we know that it hurts, or that there is a great big truck plumb in front of our eyes. Where the need for rational appraisal has to enter is in the determination that we are indeed in a position to know, and do. This need becomes urgent whenever there are grounds for fearing that we may in fact be mistaken. For to maintain any belief while dismissing, or refusing to give due weight to, reasonable and relevant objections, is to show that you are more concerned to maintain that belief than really to know whether it or some other is, after all, true.

8.8 Something similar holds too about policies and pro-grammes. Suppose we propose some policy, or support some programme, on the grounds that its implementation will lead to various shining results. Then we have to accept that, to precisely the extent that we are genuinely and sincerely devoted to those splendid objectives, we shall be eager to monitor the actual results of implementation, and ready to make or to support some appropriate change of course the moment that it emerges, if it does, that we were mistaken in thinking that these policies would in fact fulfil our original aspirations.

8.9 Suppose that we are not in this way ready and eager to learn from our mistakes. Then we make it cruelly clear that our true concern either always was, or has now become, not for the stated objectives of our policy or programme, but for something else. Perhaps our real object never was what we said it was. Or perhaps our pride and other sentiments have become involved in that programme. Or perhaps, as often happens, it itself is now for us the end and not the means. Certainly we have to suspect the true aims of all those who offer, as means to supposed goods other than their own implementation, programmes of wholesale and irreversible social change. For the more extensive and the more irreversible any such programme is, the more difficult or even impossible it must become, either to determine how far it actually is achieving its stated objectives, or to make the corrections required by discoveries of where and how far it is in fact falling short. (On such wholesale and irreversible policies see Popper (1), passim. The emphasis on the sincerity or insincerity of the ideals of those whom Popper calls ' utopian social engineers ' is, however, not his but mine.)

8.10 It should perhaps today also be pointed out—in parentheses, as it were—that democratic politicians do not promise irreversible changes. They may of course promise changes which no one will, when the time comes, want to reverse. But that is another story. What distinguishes the democratic politician here is: not that he wants to be voted in, rather than to seize power in some sort of coup; but that he is willing, in due course, to be voted out again (4.9). It is, therefore, one-eyed as well as shortsighted to welcome the French or the Italian or any other Communist Parties as belated converts to democracy; if you have no better reason than that they now promise to work to obtain power through the ballot box rather than by revolution. They still deviously make their actual intentions clear by, among other things, continuing to use the term ' democracy ' to characterize the electorally unalterable power structures of the USSR and the other Communist countries. There was by contrast something almost exhilarating about the greater frankness of Abu Zuhair Yahya, Prime Minister of Iraq in 1968: ' I came in on a tank, and only a tank will evict me ' (Quoted Luttwak, p. 146).

8.11 Someone might react to my propaganda for rationality by pointing out that the most rational of methods and approaches still provides no sure guarantee of true results; adding perhaps, as a bye-blow, that in any case rationality comprises a lot more than the capacity to discern what does and does not follow from what. This is perfectly true. History records innumerable cases of those who believed what for them it was entirely reasonable to believe, and who nevertheless were mistaken in their beliefs. Nor is there any shortage of instances in which hunches and prejudices have turned out to be right, and the contrary conclusions suggested by the best available evidence wrong.

8.12 But though the points urged in this reaction are correct, they do not upset what I am saying. First, they do not bear at all upon the original suggestion that there is a ' connection between rationality in general and personal integrity ' (8.4). Second, fallibility is one of the universal and inescapable fundamentals of the human condition. We have no choices between an option of fallibility and an option of infallibility. It is precisely our fallibility which is the best reason why we must always be open to rational criticism. Third, when and in so far as we are confronted with choices between alternative methods of inquiry, then the final judgement can only be a judgement by results. If the shaman or the soothsayer regularly and reliably comes up with predictions which are discovered to have been right, while all the economic and scientific advisers get everything wrong, then it surely becomes rational for the Minister to hire and to trust the former while firing and busting the latter. In at least one sense of ' rational ' it is paradigmatically rational to be thus guided by experience.

8.13 Yet none of this establishes, or would establish, that we can now, or could then, substitute intuition for evidence, or for argument, or for rational appraisal generally. For the very results by which the Minister, and everyone else, should judge are the discoveries of truths. So we have to be able to identify results as results—to know, that is, that the putative discoveries really are discoveries of truths—before we can say that any method of inquiry in fact is justified by results. It is here that the demand for rational appraisal arises, and with it the challenge to the sincerity of our dedication to truth. At whatever stage these questions of the justification of belief are in fact tackled, they are always logically fundamental. In terms of two ancient but still serviceable distinctions we may say that the context of justification is logically prior, albeit sometimes historically posterior, to the context of discovery.

8.14 It is because we are concerned not with mere assertion regardless of truth, nor even with mere true belief not known to be true, but with knowledge, that we are and have to be concerned with rational justification. It was with that commitment that Socrates lived, and died: 'The unexamined life is not to be endured '.

Bibliography

This is intended simply to complement the references in the text. It therefore contains no items not mentioned already. It also omits those classical works for which edition-neutral systems of reference are available: these systems have been used. This is also perhaps the place to say that in quoting from works of earlier centuries I have assimilated punctuations to the style followed in the rest of this book.

BERG, C. *Deep Analysis* (London: Allen & Unwin, 1946).

BRENAN, G. *The Spanish Labyrinth* (Cambridge: Cambridge U.P., 1943).

COCKBURN, A. and BLACKBURN, R. (Eds.) *Student Power* (Harmondsworth: Penguin, 1969).

D'ISRAELI, I. *The Quarrels of Authors* (London: John Murray, 1814).

EVANS-PRITCHARD, E. E. *Witchcraft, Oracles and Magic among the Azande* (Oxford: Oxford U.P., 1937).

FLEW, A. (1) 'A Rational Animal' in J. Smythies (Ed.) *Brain and Mind* (London: Routledge & Kegan Paul, 1965).

FLEW, A. G. N. (2) *God and Philosophy* (London, and New York: Hutchinson, and Harcourt Brace, 1966, and 1967).

FLEW, A. G. N. (3) *An Introduction to Western Philosophy* (London, and Indianapolis: Thames and Hudson, and Bobbs-Merrill, 1971).

FLEW, A. G. N. (4) *Crime or Disease?* (London, and New York: Macmillan, and Barnes and Noble, 1973).

FLEW, A. and MACINTYRE, A. C. *New Essays in Philosophical Theology* (London: S.C.M. Press, 1955).

GOSSE, E. *Father and Son* (London: Heinemann, 1907).

GOWERS, E. (1) *Plain Words* (London: H.M.S.O., 1948).

GOWERS, E. (2) *ABC of Plain Words* (London: H.M.S.O., 1951).

HALL, S. and WHANNEL, P. *The Popular Arts* (London: Hutchinson, 1964).

HUXLEY, J. *Essays of a Biologist* (Harmondsworth: Penguin, 1939).

LUTTWAK, E. *Coup d'Etat* (Harmondsworth: Penguin, 1969).

MACINTYRE, A. C. *Marcuse* (London: Fontana, 1970).

MAGEE, B. *Popper* (London: Fontana, 1973).

MARCUSE, H. *One Dimensional Man* (London: Routledge & Kegan Paul, 1964).

MORRIS, D. (1) *The Naked Ape* (London: Corgi, 1968).

MORRIS, D. (2) *The Human Zoo* (London: Cape, 1970).

ORWELL, G. (1) *Nineteen Eighty-Four* (London: Secker & Warburg, 1949).

ORWELL, G. (2) *Selected Essays* (Harmondsworth: Penguin, 1957).

PEARSON, K. *The Chances of Death* (London: Arnold, 1897).

POPPER, K. R. (1) *The Open Society and Its Enemies* (London: Routledge & Kegan Paul, Fifth edition, 1966).

POPPER, K. R. (2) *Conjectures and Refutations* (London: Routledge & Kegan Paul, 1963).

POPPER, K. R. (3) *Objective Knowledge* (Oxford: Clarendon, 1972).

RICHARDSON, K. and SPEARS, D. (Eds.) *Race, Culture and Intelligence* (Harmondsworth and Baltimore: Penguin, 1972).

ROCKWELL, J. *Fact in Fiction* (London: Routledge & Kegan Paul, 1974).

ROLT, L. T. C. *Isambard Kingdom Brunel* (Harmondsworth: Penguin, 1970).

SCHNEIDER, F. and GULLANS, C. (Eds.) *Last Letters from Stalingrad* (Toronto: Sigret, 1965).

SCHOPENHAUER, A. *The Art of Controversy*, trans. and ed. T. B. Saunders (London: Sonnenschein, 1896).

SMART, N. *The Religious Experience of Mankind* (London: Fontana, 1971).

STEPHEN, H. J. *New Commentaries on the Laws of England* (London: Butterworth, Twenty-First Edition, 1950).

STEVENSON, C. L. *Ethics and Language* (New Haven: Yale U.P., 1944).

TAYLOR, G. R. *The Doomsday Book* (Greenwich, Conn.: Fawcett, 1971).

URBAN, W. R. *Beyond Realism and Idealism* (London: Allen & Unwin, 1949).

VEBLEN, T. *The Theory of the Leisure Class* (New York and London: Macmillan, 1899).

Some Further Reading

Probably the greatest need for most people is to become less ill at ease with figures. If you think of yourself as a hopeless case, then a good first prescription would be one of two little books by Darrell Huff, illustrated by Carolyn Huff: either *How to Lie with Statistics* (Harmondsworth: Penguin, 1973); or *How to Take a Chance* (Harmondsworth: Penguin, 1970). After that or, if you are a less hard case, immediately try R. L. Meek *Figuring out Society* (London: Fontana, 1971), or D. J. Bartholomew and E. E. Bassett *Lets Look at the Figures* (Harmondsworth: Penguin, 1971), or even R. G. D. Allen *Statistics for Economists* (London: Hutchinson, Third Edition, 1966).

Another line is to develop that sort of concern with the improvement of language which is also a concern for the improvement of thought and for the cleaning and sharpening of the instruments of thought. The Orwell essays and the small books by Sir Ernest Gowers, recommended already in Chapter Five, make a good start. Some might profitably savour two brilliant and often highly entertaining essays in precisification by J. L. Austin; though these certainly are rather strong meat. 'A Plea for Excuses' and 'Other Minds' are both in his posthumous *Philosophical Papers* (Oxford: Oxford U.P., 1961).

Two outstanding earlier works in the same genre as the present book are R. H. Thouless *Straight and Crooked Thinking* (London: Hodder & Stoughton, 1930; London: Pan, 1953) and L. S. Stebbing *Thinking to Some Purpose* (Harmondsworth: Penguin, 1939). Both are much longer and more discursive, which may be advantages. But both are also, inevitably, dated in their illustrations and not perhaps in their illustrations only.

If your appetite has been whetted by the references to some key ideas of the classical philosophers, and if you want an introduction to philosophy as essentially involving argument, look at the item listed as Flew (3), above.

Index of Notions

The purpose of this index is to serve as a checklist of the ideas and distinctions specially explained and discussed in the text: it makes no attempt to include every notion which makes any appearance.